AF614765

Recent Advances on Nitrification and Denitrification

Edited by Ivan Zhu

Published in London, United Kingdom

Recent Advances on Nitrification and Denitrification
http://dx.doi.org/10.5772/intechopen.107611
Edited by Ivan Zhu

Contributors
Reshalaiti Hailili, Zelong Li, Xu Lu, Xiaokaiti Reyimu, Wenqiang Wang, Dong Li, Shuai Li, Huiping Zeng, Jie Zhang, Ikhlass Marzouk Trifi, Beyram Trifi, Lasâad Dammak, Hui-Ping Chuang, Akiyoshi Ohashi, Hideki Harada, Ivan X Zhu

First published in London, United Kingdom, 2024 by IntechOpen
IntechOpen is the global imprint of INTECHOPEN LIMITED, registered in England and Wales, registration number: 11086078, 5 Princes Gate Court, London, SW7 2QJ, United Kingdom

British Library Cataloguing-in-Publication Data
A catalogue record for this book is available from the British Library

Additional hard and PDF copies can be obtained from orders@intechopen.com

Recent Advances on Nitrification and Denitrification
Edited by Ivan Zhu
p. cm.
Print ISBN 978-1-83768-963-7
Online ISBN 978-1-83768-964-4
eBook (PDF) ISBN 978-1-83768-965-1

For EU product safety concerns:
IN TECH d.o.o., Prolaz Marije Krucifikse Kozulić 3, 51000 Rijeka, Croatia,
info@intechopen.com or visit our website at intechopen.com.

Meet the editor

Dr. Ivan Zhu has highly specialized expertise in biological wastewater treatment, membrane applications to industrial and municipal water and wastewater treatment, flocs and biofilm characterization in terms of microbial community distribution and extra-cellular polymeric substances, and membrane fouling characterization. He has applied his extensive knowledge of separation processes to the evaluation and design of water and wastewater chemical/biological treatment processes. Previously, he worked at Xylem Water Solutions, where he gained extensive experience in drinking water treatment, wastewater tertiary treatment, denitrification, biologically active filtration, ozone-enhanced biofiltration, and dissolved air flotation. Presently, Dr. Zhu is working at Evoqua Water Technologies as a senior applications engineer for integrated industrial solutions for water and wastewater treatment. He has extensive experience in the treatment of industrial wastewater from refineries, power, mining, pulp and paper, food and beverage, and more. He holds a bachelor's degree from Shanghai Jiaotong University, China, and a master's and doctoral degree from the University of Toronto, Ontario, Canada.

Contents

Preface

Wastewater treatment is a process used to convert wastewater into a treated effluent (outflowing of water to a receiving body of water) that can be returned to the water cycle with minimal impact on the environment or directly reused. Climate change, population growth, and water scarcity have contributed to a growing demand for sustainable management of water resources. With the application of nitrate-containing fertilizers, consumption of animal products, and industrial production activities, ever more ammonia and nitrate are being discharged into rivers and lakes, which may cause eutrophication and deterioration of aquatic environments.

Although there is no ammonia drinking water standard in the United States, the European community has established a maximum limit of approximately 0.5 mg/L and a guide level of 0.05 mg/L (EU Council, 1980). The maximum acceptable contamination level in drinking water is 10 mg/L nitrate nitrogen in the United States, Japan, and Korea, while the EU countries set the standard for nitrate nitrogen at 11.3 mg/L, and the World Health Organization recommends 11.3 mg/L nitrate nitrogen to protect against methemoglobinemia in bottle-fed infants. To protect aquatic ecological systems, a more stringent limit was imposed to point source dischargers into sensitive water bodies, such as Chesapeake Bay in the United States. Nitrification and denitrification are the fundamental processes in nitrogen removal in aquatic ecosystems. They play an essential role in both natural and engineered systems in terms of the nitrogen cycle.

This book reviews and updates the fundamental research and engineering experience on nitrification and denitrification. Although nitrification and denitrification are generally regarded as biological processes for the removal of ammonia, as well as nitrates and nitrites, this book covers a broad range of topics for ammonia and nitrate removal, including physical and chemical approaches.

While extensive research has been conducted on conventional wastewater treatment, this book is oriented to some interesting processes and selected applications such as photocatalytic reaction for nitrous oxide removal, autotrophic nitrogen oxidization, and the anaerobic ammonia oxidation process. Chapters in this book include:

Chapter 1: “Introductory Chapter: Ammonia Removal and Recovery”

Chapter 2: “Recent Progress and Current Status of Photocatalytic NO Removal”

Chapter 3: “Research on Partial Nitritation and Anaerobic Ammonium Oxidation Process”

Chapter 4: “The Contribution of Autotrophic Nitrogen Oxidizers to Global Nitrogen Conversion”

Chapter 5: "Removal of Nitrate and Nitrite by Donnan Dialysis: Optimization According to Doehlert Design"

This book opens possibilities for future research and innovation in the field.

Finally, during the course of editing and compiling this book, extensive support and guidance were received from Publishing Process Manager Ms Maja Bozicevic at IntechOpen The editor would like to express deep appreciation and gratefulness for her support.

Ivan Zhu
Evoqua Water Technologies, LLC,
Pittsburgh, PA, USA

Chapter 1

Introductory Chapter: Ammonia Removal and Recovery

Ivan Zhu

1. Introduction

With the application of nitrate-containing fertilizers, consumption of animal products, and industrial production activities, ever more ammonia and nitrate are being discharged into rivers and lakes, which may cause eutrophication and deterioration of aquatic environments. Traditionally, ammonia removal is achieved with biological processes such as nitrification, breakpoint chlorination, air stripping, reverse osmosis, zeolite adsorption, and so on. However, these processes either require high capital investment or are chemistry-intensive. Moreover, these processes focus on the removal instead of ammonia recovery. Ammonium nitrogen (N) is an important nutritional element in fertilizer besides phosphorus (P) and potassium (K). Recovering nutrients, instead of simply removing from wastewater, is drawing more attention to keep natural resources reliable and sustainable and minimize carbon footprint. Two processes focusing on nutrient removal and recovery have stood out in recent years and drawn attention from engineers, facility operators, and regulators.

2. Gas-permeable membrane for ammonia recovery

In the application of a gas-permeable microporous membrane, the wastewater stream is first adjusted to a pH value of at least 9.5 (**Figure 1**). And then, the stream passes through one side of the membrane and dissociates ammonium from water, and ammonia penetrates through the membrane; a dilute acid solution is circulated on the other side of the membrane, and sequesters ammonia to form ammonium sulfate (**Figure 2**).

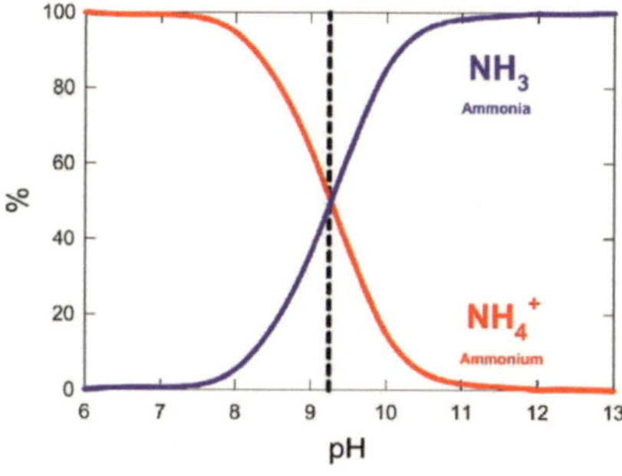

Figure 1.
Ammonium speciation in water at different pH [1].

IntechOpen

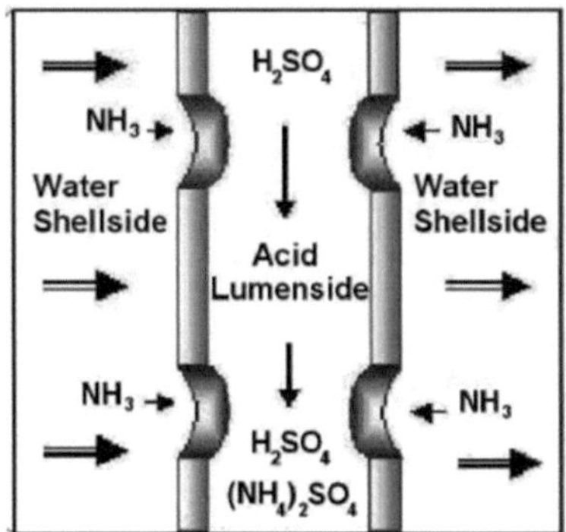

Figure 2.
Schematic diagram for gas-permeable membrane for ammonia recovery (from [2]).

Ammonium sulfate solution can be used as a by-product for fertilizers or other purposes. The hydrophobic hollow fiber membrane can be used as a medium to separate aqueous phases because it is not inherently selective between permeating species. The driving force of mass transfer is the concentration difference between the two sides of the membrane. On the two sides of the membrane, pH values are distinctively different. On the wastewater side, the pH is at least 9.5 or higher, and on the dilute acid side, the pH is 2 or lower.

This process was applied in full-scale systems in manufacturing facilities [2, 3]. It was found that removal rate was achieved as high as 95 to 97%. The two important operating parameters are wastewater pH and temperature. With an increasing pH value of the wastewater stream, more ammonium species is converted from ammonium to ammonia; the mass transfer is thereby enhanced, and ammonia removal efficiencies are further improved. Since Henry's law constant increases at an elevated temperature, and favors the gas phase concentrations (**Figure 3**), temperature will also affect the rate of transfer from the liquid to the gas phase, with faster rate at higher temperatures. Therefore, increased temperature will improve the ammonia removal in this application. On the dilute acid side, it is circulated counter current,

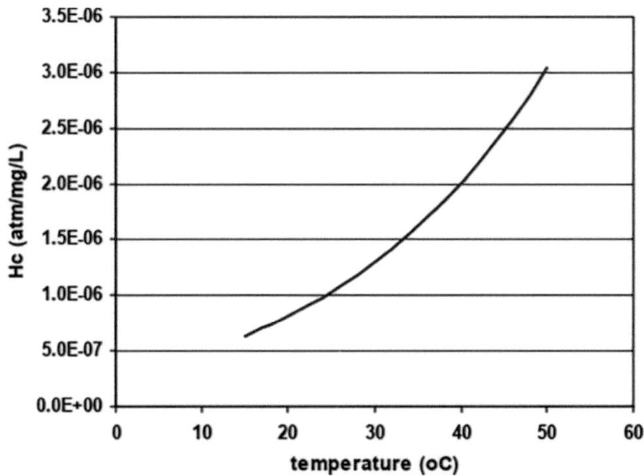

Figure 3.
Effect of temperature on Henry's law constant for ammonia [4].

and pH is maintained 2 or less by supplementing acid, until ammonia sulfate reaches a certain level. Typically, ammonium sulfate concentrations can reach to 25 to 30%. Because the acid stream is not in contact with the wastewater stream, high-quality ammonia sulfate solution can be obtained.

This process has significant advantages such as low capital investment, small footprint, lower energy cost, and recovery of a valuable by-product as ammonium sulfate.

Currently, 3 M and duPont are promoting their own Degasification Membrane Modules for the application of ammonia removal and recovery.

3. Struvite precipitation for nutrient recovery

Struvite is a crystal compound, consisting of magnesium, phosphorus, and nitrogen. It naturally forms in many parts of a wastewater treatment plant, such as anaerobic digesters, aerobic sludge digesters, digestate pipes and pumps and valves, plant feed pumps , sludge-holding tanks or thickeners, centrifuges, outfall pipes, centrate pipes, and so forth. It creates scales and may also cause process disruptions. Its chemical formula is $NH_4MgPO_4.6H_2O$. Stoichiometric reaction is shown below:

$$Mg^{2+} + NH^{4+} + PO43^{-} + 6H_2O = MgNH_4PO_4 : 6H_2O \tag{1}$$

This compound is slightly soluble in water. When spread in agriculture fields, it slowly releases nitrogen and phosphorus, which are important nutritional elements for plant growth.

Digested sludge usually consists of high concentration of ammonia and phosphate, because bacteria decompose and release ammonia nitrogen and phosphorus during aerobic or anaerobic digestion. Animal wastes, such as swine waste, also consists of high concentrations of ammonia and phosphorus. These wastewater streams may cause concerns of scale formation and process disruptions and yet present opportunity for nutrient removal and recovery if properly managed.

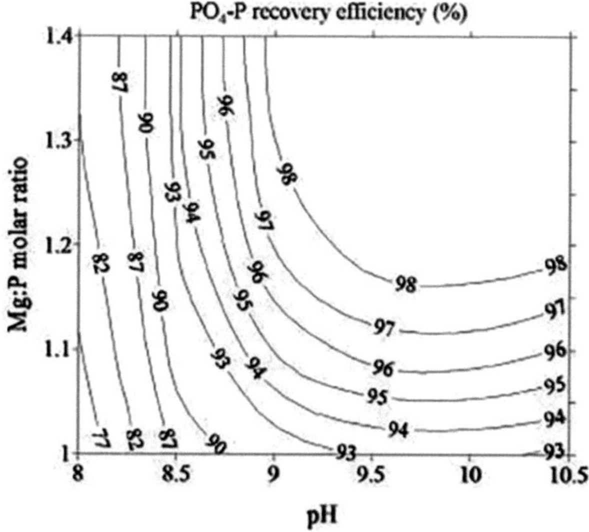

Figure 4.
Influence of pH and Mg/P molar ratio on P recovery [5].

Under proper temperature and pH, struvite forms when the concentrations of ammonium, phosphate, and magnesium exceed the solubility product, and struvite precipitates in a molar ratio of 1:1:1 of ammonium, phosphate, and magnesium.

It was summarized that the preferred pH rage is 8 to 9 and a high Mg/P ratio drives a higher efficiency of struvite conversion [6]. The presence of calcium, however, negatively affects the formation of struvite. **Figure 4** shows the influence of pH and Mg/P ratio on the struvite formation efficiency.

Ostara's Peral® process, which recovers struvite through controlled precipitation in a fluidized bed from digestate, was successfully applied in plants in North America and Europe. It removes and recovers nutrients (ammonia and phosphorus) from digestate and produces high premium-crystal-form struvite, which, in turn, is recycled back for agricultural use as a fertilizer [7]. Using a fluidized bed reactor for struvite precipitation, it was found the payback period is less than 10 years in Budds Farm wastewater treatment plant in England [8].

Author details

Ivan Zhu
Evoqua Water Technologies, LLC, Pittsburgh, PA, USA

*Address all correspondence to: ivan.zhu@evoqua.com

References

[1] Sawyer CN, McCarty PL. Chemistry for Environmental Engineering. 3rd ed. NY: McGraw-Hill Book Company; 1978. 532 p

[2] Stasiak M, Ulbricht M, Schneider J, Munos J, Sengupta A, Kitteringham B, et al. New Fab Technical Journal. 2011;**July**:90-94

[3] Ulbricht M, Schneider J, Stasiak M, Sengupta A. Chemie Ingenieur Technik. 2013;**85**(8):1259-1262

[4] Liehr SK, Classen JJ, Humenik FJ, Baird C, Rice M. ASABE Annual International Meeting, Portland OR. St. Joseph, MI: ASABE; 9-12 July 2006

[5] Huang H, Guo G, Zhang P, Zhang D, Liu J, Tang S. Feasibility of physicochemical recovery of nutrients from swine wastewater: Evaluation of three kinds of magnesium sources. Journal of the Taiwan Institute of Chemical Engineers. 2017;**70**:209-218

[6] Enyemadze I, Momade FWY, Oduro-Kwarteng S, Essandoh H. Phosphorus recovery by struvite precipitation: A review of the impact of calcium on struvite quality. Journal of Water, Sanitation and Hygiene for Development. 2021;**11**(5):706-718

[7] Gysin A, Lycke D, Wirtel S. Chapter 19 the pearl® and WASSTRIP® processes (Canada). In: Phosphorus: Polluter and Resource of the Future. London, UK: IWA Publishing; 2018. pp. 359-365

[8] Achilleos P, Roberts KR, Williams ID. Struvite precipitation within wastewater treatment: A problem or a circular economy opportunity? Heliyon. 2022;**8**:e09862

Chapter 2

Recent Progress and Current Status of Photocatalytic NO Removal

Reshalaiti Hailili, Zelong Li, Xu Lu and Xiaokaiti Reyimu

Abstract

Air pollution has become a globally prominent environmental problem in which nitrogen oxide (NO_x, 95% NO and NO_2) has been considered as one of the most serious harmful gaseous pollutants that can cause haze, photochemical smog, and acid rain. Exposure to NO (~ppb) harms human health with a risk of respiratory and cardiopulmonary diseases. As such, much attention is focused on the throughout removal, effective control, and precise monitoring of NO, especially for those of NO with low concentration (~ppb). Semiconductor-based photocatalysis is a practical approach for pollutant treatments, especially for low concentrations but highly toxic ones, for example, NO (~ppb) removal in indoor and outdoor atmospheres. This work aims to introduce the main process, methods and summarize the critical scientific issues during the photocatalytic NO treatment and review the latest progress in semiconducting materials. This work also surveys the newly emerged photocatalysts such as metal oxides, Bi-based semiconductors, including $Bi_2O_2CO_3$, BiOX (X = Cl, Br, and I), Bi-metal-based defective photocatalysts, and other Bi-based catalysts with well-defined surface/interface characters for the complete NO removal, specific conversion mechanisms and controlling the generation of the toxic intermediate (NO_2) is highlighted. The challenges/bottlenecks of the practical applications in the field are also highlighted at the end.

Keywords: photocatalysis, nitrogen oxides (NO_x), visible light, microstructure, surface defects, selectivity

1. Introduction

According to the World Health Organization (WHO) health report in 2021, air pollution caused about 7 million deaths worldwide, of which 4.2 million die prematurely every year by outdoor pollution, and 3.8 million by indoor pollution (**Figure 1a**) [1]. Air pollutants also affect local plants and animals' survival and cause environmental issues. Nitrogen oxides (NO_x, mainly NO + NO_2), primarily emitted from power stations, factories, and automobiles are regarded as the major source of atmospheric contaminations, which remarkably influence the tropospheric chemistry and become the main cause of the greenhouse effects, acid rain, photochemical smog, and $PM_{2.5}$ [2]. Generally, with an unpaired electron $\{(\sigma 2s)^2(\sigma^* 2s)^2(\sigma 2px)^2(\pi 2py)^2(\pi 2pz)^2(\pi^* 2py)^1\}$, NO is chemically active and reacts with oxygen readily to generate NO_2.

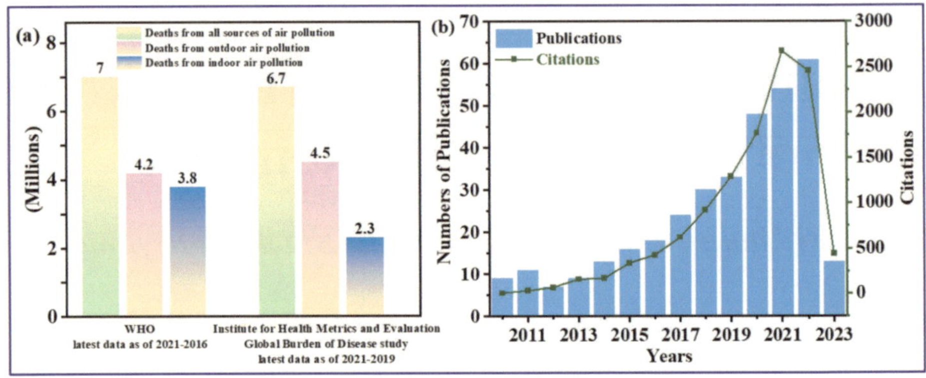

Figure 1.
(a) WHO 2021 report on the correlations between mortality and environmental pollution; (b) numbers and citations of publications about photocatalytic NO removal since 2010.

However, the reaction activity significantly descends when the concentration of NO is lower than the ppm level and it can exist stably in the air for a long time. Long-term exposure of NO even at such a low level still causes serious respiratory diseases, such as lung cancer, emphysema, and asthma. Moreover, the concentration of NO in the indoor environment is up to 200 ppb, because of cooking or smoking activities. Concerning such severe harms, strict legislation and policies have been enacted to control the emissions of NO in many countries. For instance, in 2004, the USA claimed the maximum emission rate of NO is required not to exceed 553.5 mg m^{-3} for tangentially fired boilers [3]. The limit of NO from light-duty vehicles is set at 35 mg km^{-1} in China (GB 18352.6-2016) [4]. According to the WHO guideline, the lowest NO_x concentration emission threshold under the current was set at 40 μg m^{-3} [5]. Thus, it is urgent to develop environmentally friendly, effective methods and evolution of technical measurements that could control emissions, decrease the concentrations, and reduce the harm of nitrogen oxides, for example, NO at the level of hundreds of ppb in indoor circumstances. Conventionally, methods such as physical/chemical adsorption, advanced wet oxidation, and post-combustion reduction technologies such as selective catalytic reduction with NH_3 (SCR-NH_3) and hydrocarbons (SCR-HC), are used for minimizing the NO emission [6]. The chemical adsorption method refers to water or aqueous solutions of acids, bases, and salts to absorb the nitrogen oxides in the exhaust gas so that the exhaust gas can be purified. This method has low investment in equipment and low operating costs. However, the absorption efficiency is not high, and the purification effect is poor for the exhaust gas containing more NO, and it is not suitable for treating the exhaust gas with large volumes. SCR is also regarded as an efficient method to reduce the values of NO. For instance, Dong et al. reported the "anchoring (NH_3/He plasma)-reduction (O_2/N_2 plasma)" plasma-catalytic circular system for the NO reduction at room temperature in the presence of O_2 and revealed that NO was firstly oxidized to NO_2 by plasma, then NO_2 experienced part-disproportionation reactions during adsorption on oxygen vacancies, producing NO, N_2O, N_2, and nitrate [7]. However, these approaches require high temperatures, special handling systems, and sophisticated equipment to avoid NH_3 slip, reducing agents, or cocatalysts, meanwhile suffering from high costs and yields of more toxic byproducts. Moreover, the above techniques are no longer suitable for the complete removal of dilute air pollutants, for example, gaseous NO at the ppb level. Utilizing endless solar light as the driving force, semiconductor-based photocatalysis has received considerable attention to remove the atmospheric gaseous hazards from the atmosphere, especially for the ones in low concentration but highly toxic, for example, NO (~ppb), in an economically attractive and environmentally friendly manner [8–10]. In view of the current state of NO removal, photocatalytic oxidation is the most studied way to reduce NO concentration from the air. As shown in **Figure 1b**, the number of papers on photocatalytic NO removal and their citations has rapidly increased in the past 13 years.

2. Methods, basic principles, and main scientific issues for photocatalytic NO removal

There are three commonly accepted methods for the removal of NO_x by photocatalysis: photodecomposition, photo-selective catalytic reduction (Photo-SCR), and photooxidation. Both direct and photo-SCR decomposition can be classified as the typical reduction pathways, while NO photooxidation belongs to the photoinduced oxidation methods. Through these approaches, the concentrations of NO in the atmosphere can successfully be decreased or will be completely removed by their conversion into the other N, O contained species such as nontoxic harmless N_2, O_2, or intermediates of NO_2, N_2O_4, or even the deeper oxidation products of HNO_3. Evidently, direct NO decomposition into nontoxic and harmless products of stoichiometric N_2 and O_2 offers one of the most ideal routes for the NO treatment, in which no additional reductants are required, and the side reactions are hence minimized. Although this reduction process is exothermic and thermodynamically favored, several kinetically sluggish steps such as the breakage of N=O bonds (153.3 kJ mol^{-1}) and subsequent reconstruction of structurally stable N≡N triple bonds (940.95 kJ mol^{-1}) are involved, both of which require to overcome the huge energetic barriers [8, 11]. The Photo-SCR process occurs on a photocatalyst surface and involves the reduction of NO_x in the presence of reducing agents such as NH_3 and hydrocarbons under light irradiation. However, the trigger of this multi-step reaction requires a temperature higher than 500°C, and undesired compounds intermediate byproducts (e.g., N_2O) are released instead of the stoichiometric formation of harmless N_2 and O_2, affecting the reaction efficiency as well as selectivity [12]. Compared with traditional NO removal methods, more selective and nonselective products, such as nitric acid or nitrates, are generated *via* NO photooxidation and thus reduce its harm. However, these oxidative species will block the surface-active sites of photocatalysts, restrict further redox reactions and consequently reduce the reaction efficiency. Hence, these species are required to be removed rapidly from the surface of the photocatalyst to avoid catalysts deactivation. The main methods/ aims, advantages, and disadvantages of photocatalytic NO removal are summarized in **Figure 2**. Obviously, the photocatalytic oxidation of dilute NO to NO_3^- with well-defined photocatalysts is preferred as the ideal pathway that is low-cost, environmentally friendly, and suitable for the removal of low concentrations of NO.

The main principles of photocatalytic NO removal as followings: (i) formations of electron-hole pairs: when the photocatalyst is irradiated with photons with the same or higher energy than its band gap, the electrons are excited from the valence band to the conduction band of semiconductor, and leave same amounts of holes in the VB, hence, resulting in the formations of electron-hole pairs; (ii) charge carrier separation and migration: the photoinduced electrons and holes will migrate to the surface, where photocatalytic redox reaction will further be initiated; partial electron and holes

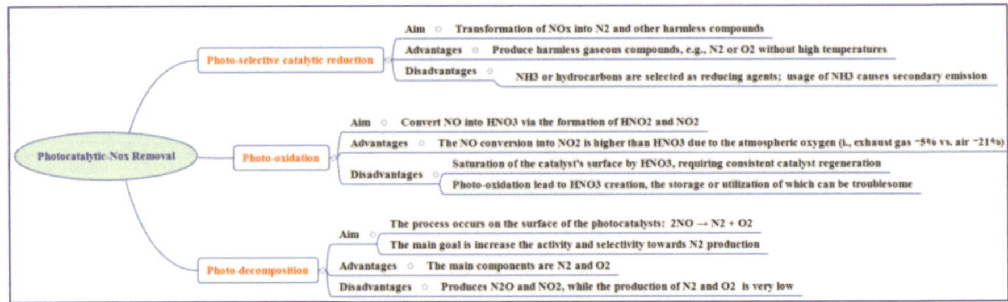

Figure 2.
A summary of advantages and disadvantages of photocatalytic NO removal.

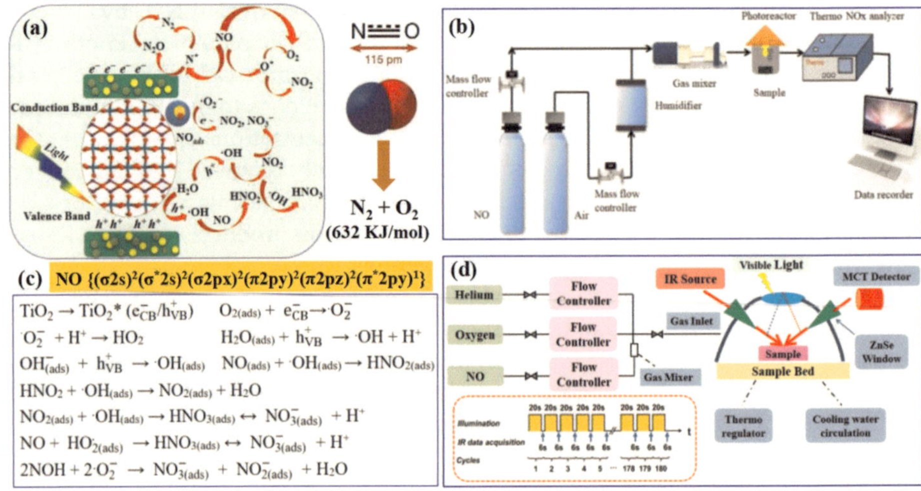

Figure 3.
(a, c) Schematic diagram of the photocatalytic NO conversion process, (b) reaction setup, and (d) in situ *DRIFTS tests for photocatalytic removal of NO under room temperature.*

migrate at the bulk or surface of the photocatalysts, while some of them will lead to the emission of light or heat; (iii) surface redox reaction: the photogenerated electrons are trapped by surface adsorbed molecular oxygen to generate various reactive oxygen species (ROS), including hydrogen peroxide (H_2O_2), single line state oxygen ($^1O_2$), superoxide radicals ($\cdot O_2^-$), and hydroxyl radicals ($\cdot OH$), which react with NO through different pathways to generate oxidation species such as NO^+, N_2O_4, NO_2^-, or be deeply oxidized into NO_3^-; Under oxygen free system, the photoinduced electrons would generate assorted intermediates, such as N_2O^* or $^*NO_2$., and subsequently converted into completely harmless N_2 or O_2, and reduce the concentrations of NO (**Figure 3a**). Among them, $\cdot O_2^-$ can completely convert NO to the final nitrate ($NO + \cdot O_2^- \rightarrow NO_3^-$), while $^1O_2$ will oxidize NO to unwanted NO_2 ($NO + {}^1O_2 \rightarrow NO_2$). On the other hand, HNO_2 is the main oxidation species at the initial state and is closely adhered to the surfaces of the catalysts. Note that the main intermediate NO_2 possesses nearly 9-fold higher toxicity than that of dilute gaseous NO and should be minimized or restricted during the photocatalytic NO conversion [13]. It was reported that the adsorbed HNO_2 will further dissociate to NO_2^- with further light illuminations, and subsequently oxidized into NO_3^-, which can be washed away from the catalyst's surfaces (**Figure 3b**). Unfortunately, the product nitrate does not desorb and adheres to the catalyst's surface and occupies the active surface sites of the catalysts to reduce its activity, resulting in massive deactivation of the catalyst's powder, and subsequently blocking further conversion. **Figure 3c** displays the main reaction setup for photocatalytic NO removal, in which the catalysts powders were exposed to light, and the effluent NO and NO_x ($NO + NO_2$) concentrations were continuously recorded using the online chemiluminescence NO_x analyzer. The concentration changes of the intermediate NO_2 were simultaneously calculated from concentration gaps between NO and NO_x during the tests. The final conversion products are analyzed by online separation methods, for example, NO_3^- can be detected by ionic chromatography, while the decomposition products of N_2 and O_2 are detected by gas chromatography analyses. Generally, the NO removal process includes the initial state, transition state, and final step, during which the conversion products are closely related to the reaction conditions such as continuous reactor, NO concentration, flow rate, temperature, water (relative humidity), catalysts loading, reactor sizes, air flow and light parameters (wavelength or intensity), detected conversion products and so on [14]. With tuning the above reaction conditions, the main conversion products of NO can be controlled, and reverse reactions can be prevented to enhance reaction efficiencies. Given upon assorted conversion products during the photocatalytic NO removal, the reaction displays certain selectivity toward specific products such as selectivity for NO_3^-, N_2, or NO_2. For the typical photocatalytic NO conversions, the corresponding removal efficiency, NO_x conversions (%), N_2 and NO_2 selectivity are calculated by the following equation [15–17]:

$$\text{NO conversion } (\eta_{NO}, \%) = \left(1 - \frac{[NO]_{in}}{[NO]_{out}}\right) \times 100\% \tag{1}$$

$$NO_x \text{ conversion } (\eta_{NOx}, \%) = \left(1 - \frac{[NOX]_{out}}{[NOX]_{in}}\right) \times 100\% \tag{2}$$

$$N_2 \text{ selectivity } (S_{N_2}, \%) = \frac{[N_2]_{out}}{[N_2]_{out} - [NO_2]_{out}} \times 100\% \tag{3}$$

$$NO_2 \text{ selectivity } (S_{NO_2}, \%) = \frac{[NO_2]_{out}}{[NO]_{in} - [NO]_{out}} \times 100\% \tag{4}$$

where $[NO]_{in}$, $[NO]_{out}$ is the initial and final concentrations of NO, while $[NO_x]_{in}$, $[NO]_{out}$ and $[NO_2]_{out}$ represent the initial and final concentrations of NO_x and NO_2 in ppb level, respectively.

The main NO oxidation products NO_2, NO_3^-, N_2O, and N_2O_4 etc., are strongly dependent on the surface characters of semiconducting materials and the assorted NO oxidation products can be obtained. The time-dependent generation of intermediates and products of NO on the surface of catalysts powders are monitored by the *in situ* diffuse reflectance infrared Fourier transform spectroscopy (DRIFTS) measurements (**Figure 3d**).

3. Latest progress in photocatalytic NO removal

Many successful state-of-the-art photocatalysts have been explored for NO removal since the photocatalytic process can be conducted under ambient conditions without the addition of extra redox reagents, which is specifically applicable to indoor circumstances. It can be found that great progress has actively been pursued in many research groups and successful state-of-the-art catalysts, for example, TiO_2 and TiO_2-based oxides, Bi-based compounds BiOX (X=Cl, Br, and I), metal-free catalysts, for example, g-C_3N_4 and some plasmonic metals (Au, Ag, and Bi), etc., have been developed for photocatalytic NO removal. These informative and fundamentally important studies provided encouraging results for commercial applications of photocatalysts for air cleaning and NO removal. However, the photoconversion efficiency remains low (apparent quantum yield 2.5%) due to hard control energy band gaps causing inevitable charge recombination and most of these reactions involved releasing more toxic gas NO_2 or the photocatalysts have been suffering from deactivation. In view of the current state of photocatalytic NO removal, the development of effective and green NO control technologies is of great significance to controlling air quality. It can be seen from the state-of-the-art that the most widely studied semiconductor photocatalysts for photocatalytic oxidation and reduction to remove low-concentration NO are mainly TiO_2 and metal-based materials, bismuth-based materials, graphite phase carbon nitride (g-C_3N_4)-based materials and other heterojunction systems. We have made a specific summary of these photocatalysts used for the removal of low-concentration NO at room temperature.

3.1 TiO_2 and other metal oxide-based materials

To remove dilute NO from the air by transforming NO into HNO_3 to form HNO_2 and NO_2 species upon light illumination, the previous investigations focused on the famous star catalyst TiO_2 and TiO_2-based photocatalysts. Wang et al. proposed the additional reactions, in which NO conversions over TiO_2 by the process of charge carrier generation, the trapping of a hole and an electron to produce "active" hydroxyl and oxygen radicals, and further oxidized to final product by $NO \rightarrow HNO_2 \rightarrow NO_2 \rightarrow HNO_3$ pathway [18]. Following this, Hunger et al. investigated kinetic models for photo-removal of NO_x over TiO_2 under UV light and proposed that the kinetic parameters of NO oxidation were determined by the concentration of NO and NO_2, flow rate, relative humidity, and light parameters [19]. As evidenced, the NO conversion products over TiO_2 hugely influence the reaction selectivity [20]. In this regard, the blue TiO_2 with highly abundant oxygen deficiency was investigated and displayed the highest selectivity of 99% under visible-light irradiation for the photocatalytic NO oxidation to nitrate without NO_2 yields. It was claimed that oxygen defects could not only activate molecule O_2 to generate $\cdot O_2^-$ that facilitated the selective NO transformation toward nitrate under ambient conditions but also could annihilate the photogenerated holes to further inhibit the byproduct NO_2 formation [16]. The elusive NO_3^- conversion mechanism over P25 is revealed by Dong et al., and it was revealed that the N—O bond in surface NO_3^- could be activated by NO molecules, arising from the significant overlap of the 2p orbitals between the N in NO and the O in NO_3^-. Then, photogenerated electrons (e^-) captured by NO drive the transformation of NO_3^- under light irradiation *via* the $NO_3^- + NO^- \rightarrow 2NO_2^-$ route. Additionally, although photogenerated holes (h^+) and $\cdot OH$ radicals could oxidize NO into NO_3^-, the rate of production of NO_3^- is much slower than that of photochemical transformation by NO^-. Hence, the photochemical transformation of surface NO_3^- can be solved only by preventing the formation of NO^- during photocatalytic NO oxidation (**Figure 4**) [21]. The nitrate photolysis on photoactive TiO_2 particles in the presence of SO_2 was investigated through density functional theory (DFT) simulation and *in situ* DRIFTS analysis. It was found that the nitrate was oxidized to $^*NO_3$ radicals

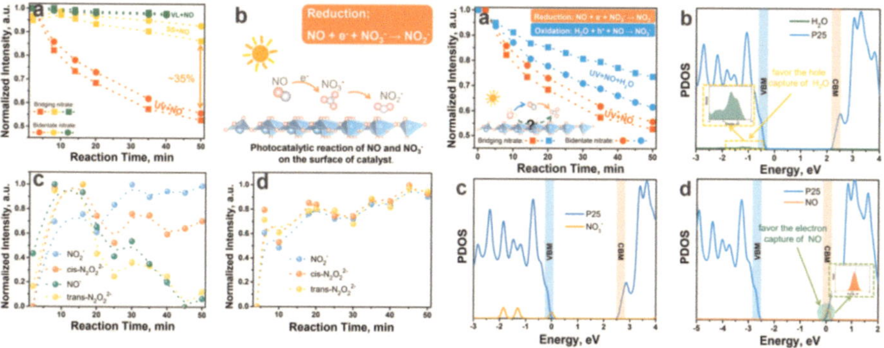

Figure 4.
Experimental and theoretical investigations of surface NO_3^- reduction by NO and e^- with light illumination on the surfaces of P25. Reproduced with permission from Ref. [21], Copyright © 2021, American Chemical Society.

by the holes generated on the surface of TiO_2, followed by reactive nitrogen species generation *via* NO_3 radicals' reduction by the photoinduced electrons. It was indicated that photogenerated h^+ plays a key role in nitrate photolysis on photoactive mineral dusts with or without the coexistence of SO_2, providing new insights into the source of NO_x and HONO in complex air-polluted areas during the daytime [22].

In addition, other metal oxide-based semiconductors have been investigated to improve NO conversions as well as reaction selectivity. Lei et al. have investigated the defective α-Bi_2O_3 and β-Bi_2O_3 to explore the synergistic effects of crystal structure and Vo on photocatalysis and reported the highly efficient photocatalytic NO removal. With surface defects, the photocatalytic NO removal over β-Bi_2O_3 was increased from 25.2 to 52.0%, while α-Bi_2O_3 indicated NO enhancement just from 7.3 to 20.1%. The improved NO performances were attributed to Vo, which could synergistically regulate the electron transfer pathway [23]. To tackle the bottlenecks of the sluggish carrier separation, catalysts deactivation, and incomplete oxidation during the photocatalytic NO treatments, Hailili et al. fabricated a series of ZnO nanostructures with gradient Vo and investigated their NO oxidations. Results showed that with higher Vo on the unusual nonpolar facets, Vo-rich ZnO exhibited 5.43 and 1.63 times enhanced NO removal with fewer toxic product NO_2 formations than its counterparts pristine and Vo-poor ZnO due to the promoted carrier separation, massive productions of $\cdot O_2^-$ radicals from the molecular oxygen activation, and effective adsorptions of small molecules (O_2, H_2O, and NO) on the defective surface [15]. Continuing their interesting studies, they investigate the influences of defect-induced surface interface for NO removal. It was found that with well-positioned band edges, defect-associated carrier separations, and strengthened surface-interface reaction, ZnO displayed 4.16 folds enhanced efficiency and 2.76 times decreased NO_2 yields, indicating the significance of surface-interface regulations and surface defect controlling [24].

3.2 Bismuth-based materials

The synthesis of highly efficient photocatalysts and the revealing of the interfacial reaction mechanism are two major prerequisites for the commercial application of photocatalytic technology. Among all the studied systems, Bismuth-based photocatalysts have received extensive attention due to their unique layered structure, electronic configurations and photoelectric properties. The layered structure polarizes the internal atomic orbitals, leading to the formation of the internal electric field, which in turn promotes the separation of photogenerated electrons and holes. Moreover, this unique two-dimensional layered structure has abundant active sites and an easily adjustable band gap. Many bismuth-based photocatalysts with excellent photocatalytic activity have been explored for the removal of dilute NO (~ppb) from the environment. However, limited visible-light absorptions, rapid carrier recombination, and ambiguous reaction mechanism with uncleared NO conversion products hindered their practical implementations. Methods such as surface defect engineering and metal doping, especially with Bi-metal incorporations turn out to be the applicable way to tackle such bottlenecks, in which the presence of surface Vo can affect all three basic steps of photocatalytic NO oxidation: i) photon energy absorption by semiconductor photocatalysts to generate photogenerated carriers, ii) separation and transfer of photogenerated electrons and photogenerated holes, and iii) surface reactions between electron holes and substrate molecules (NO, O_2, H_2O, etc.). Herein, we present several types of mostly investigated Bi-based photocatalysts that are utilized for the removal of NO by oxidation reaction.

3.2.1 $Bi_2O_2CO_3$-based materials

With low toxicity, controllable structure, and facile preparation, $Bi_2O_2CO_3$ with non-centrosymmetric crystal structure containing unique $[Bi_2O_2]^{2+}$ and CO_3^{2-} layers have been investigated as promising photocatalysts in the field of environmental photocatalysis, especially in NO purification. For photocatalytic NO removal, a significant challenge is to achieve catalytic stability while maintaining high conversion efficiency. N-doped $Bi_2O_2CO_3$ with (001) and (110) exposed facets were synthesized by tuning the pH in the hydrothermal processes and displayed crystal facet dependent NO conversion. Results showed that the N-doped $Bi_2O_2CO_3$ with (110) exposed facets are more beneficial for the activation and adsorption of NO molecules, and further reduce the activation energy, thus promoting the selective conversion of NO to the target products to inhibit the formations of toxic intermediate NO_2 [25]. The B-doped $Bi_2O_2CO_3$ hierarchical microspheres exhibited remarkably enhanced visible light NO conversion to the nitrates *via* important NO_2^+ intermediates due to the massive production of active specie, NO molecule activation and subsequently promoted charge carrier separations [26]. In further investigation, La-doped $Bi_2O_2CO_3$ was studied to simultaneously improve the photocatalytic NO conversion efficiency and selectivity to target products (NO_2^-/NO_3^-). The experimental and theoretical simulations indicated that the O_2 and NO could exchange electrons with localized excess electronic and get activated to produce more active species [27]. To highly maintain NO removal efficiency and NO_2 production, a $Bi_2O_2CO_3$/β-Bi_2O_3 heterostructure is developed and further decorated with graphene quantum dots. By construction of such an efficient interfacial charge transport channel, the charge carrier separation is hugely promoted in this heterojunction displaying high efficiency and stable visible light NO-photooxidation [28]. Zhu et al. investigated the influences of the two different crystallographic positions of oxygen atoms in the $[Bi_2O_2]^{2+}$ layer of $Bi_2O_2CO_3$ for reactive oxygen species generations as well as NO oxidations. Results showed that samples showed 50.0 and 41.6% NO removal efficiencies with generations of 15.6 and 16.54 ppb NO_2, respectively. The formation mechanism of the position-manipulated Vos and the mechanism of photooxidative NO removal over the BOC were well-disclosed (**Figure 5**) [29]. Lee et al. reported the p-n type of $Bi_2O_2CO_3/ZnFe_2O_4$ heterojunction for the removal of NO and obtained improved NO conversion due to the massive production of $\cdot O_2^-$ radicals and carrier separation induced from an

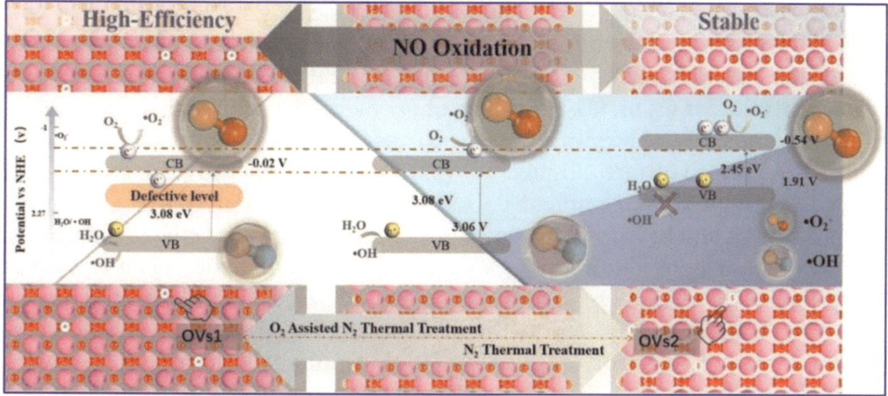

Figure 5.
Proposed schematic diagram for the migration and separation of electron-hole pairs and photocatalytic process over surface Vos-induced $Bi_2O_3/Bi_2O_2CO_3$ heterostructure photocatalyst Reproduced with permission from Ref. [29], Copyright © 2021, American Chemical Society.

internal electric field at the interface of the catalysts [30]. Furthermore, Vo-induced heterojunctions $Bi_2O_3/Bi_2O_2CO_3$ exhibited superior gas adsorption and improved NO oxidation due to the effective production of $\cdot OH$ and $\cdot O_2^-$ radicals, and the presence of surface defects changed the NO removal pathways [31]. A heterojunction $Bi_2Mo_3O_{12}$@$Bi_2O_2CO_3$ was designed and favorably synthesized in a hydrothermal way and further employed for NO removal. Results showed that promote NO oxidation was achieved over this heterojunction, and reaction process was monitored by *in situ* DRFTS, which revealed the detailed NO adsorption and conversion process as final product (NO_3^-) *via* several important intermediate products (NO^-, NO_2^-, and NO_2), all raised from the effective carrier separation, migration, and conversion of photoinduced electron-hole pairs [32]. The hybrid of two-dimensional/two-dimensional (2D/2D) Mo-g-C_3N_4 (Mo-CN) and $Bi_2O_2CO_3$ (BOC) materials displayed 45% NO removal since the interfaces led to stronger interfacial interaction and Vo due to the introduction of Mo atom in contrast with bare graphitic carbon nitride (g-C_3N_4; CN) and BOC [33].

3.2.2 BiOX (X = Cl, Br, and I)-based materials

Bismuth halide oxide BiOX (X = Cl, Br, and I) is a sillén-structured bismuth-based semiconductor material consisting of two interlaced layers of halogen atoms and $[Bi_2O_2]^{2+}$ layers. Controlling and blocking the generation of highly toxic intermediates through regulating the reactive species during the NO oxidation was investigated over these kinds of Bi-based photocatalysts. For instance, Vo containing BiOCl was investigated and demonstrated enhanced NO removal from 5.6 to 36.4% as well as obviously inhabited NO_2 generation was found under visible-light irradiation. Results revealed that Vo on the surface of BiOCl speeds the trapping and transfer of localized electrons to activate the O_2 to produce $\cdot O_2^-$ radicals, which avoid NO_2 formation, resulting in complete oxidation of NO ($NO + O_2^- \rightarrow NO_3^-$) [34]. Dong et al. revealed the dynamic evolution of surface defects in BiOCl during the gas-solid photocatalytic reaction at the electronic level by *in situ* electron paramagnetic resonance (EPR) technology. It was disclosed that the *in situ*-generated surface defects are the real active sites and can effectively activate the reactant molecules *via* directional single-electron transfer [35]. Zhang et al. reported the photocatalytic NO conversion with 99% selectivity using a defective BiOCl with (001) surface. Mechanism investigation disclosed that Vo on its prototypical (001) surface of BiOCl allows the selective and efficient activation of O_2 to $\cdot O_2^-$ in different geometric structures that thermodynamically suppressed the terminal end-on O_2^- associated NO_2 emission and selectively oxidized NO to nitrate (**Figure 6a**) [17]. In their further investigations, they developed the two-electron-trapped V_O of BiOCl, in which a prototypical F center (V_O''), is a superb site to confine O_2 toward efficient and selective NO oxidation to nitrate. Upon solar light illumination, V_O'' completes NO oxidation *via* a two-electron charging ($V_O' + O_2 \rightarrow V_O''\text{-}O_2^{2-}$) and subsequent one-electron de-charging process ($V_O''\text{-}O_2^{2-} + NO \rightarrow V_O\text{-}NO_3^- + e^-$). The back-donated electron is re-trapped by V_O to produce a new single-electron-trapped Vo (V_O'), simultaneously triggering a second round of NO oxidation ($V_O'\text{-}O_2 + NO \rightarrow V_O\text{-}NO_3^-$) (**Figure 6**) [9]. Yuan et al. reported that Mn_3O_4/BiOCl achieves about 75% of NO removal within 10 min, and not only exhibited superior inhibition for NO_2 under light irradiation, but the activities gradually decreased due to the accumulation of products. Moreover, the NO removal efficiency increased with the addition of 5–10% H_2O gas, meanwhile displayed remarkably reduced NO_2 inhibition [36].

The simultaneous incorporations of Ba and Vo into BiOBr nanosheets were investigated and displayed ~10 times enhanced NO removal than the pristine BiOBr under visible-light irradiation [37]. Utilizing the layered structure stacked by the Bi-O layer and halogen anions layer, Zhang et al. reported the novel $BiOCl_xBr_{1-x}$, ($0 \leq x \leq 1$) catalysts and demonstrated that the mixture anions products have a preferable reaction thermodynamics for NO removal with the highest efficiency of 60%, especially the $BiOCl_xBr_{1-x-3:1}$. It was presented that the improved activity was not linear dependence with light harvesting and charge conversions but was mainly decided to the optimized reaction thermodynamics, as $BiOCl_xBr_{1-x-3:1}$ possess the lowest thermodynamic barrier for NO oxidation [38]. The BiOBr/SnO_2 heterojunction exhibited more efficient NO oxidation but also inhibited the production of toxic NO_2 due to the effective carrier transportation and separation under the influence of internal electric field [39]. To reveal true reaction mechanisms, Zhang et al. investigated the NO removal over surface boronized BiOBr and proposed that the robust excitonic effect of BiOBr nanosheets, which is prototypical for $^1O_2$ production to partially oxidize NO into a more toxic NO_2, can be weakened by surface boronizing *via* inducing a

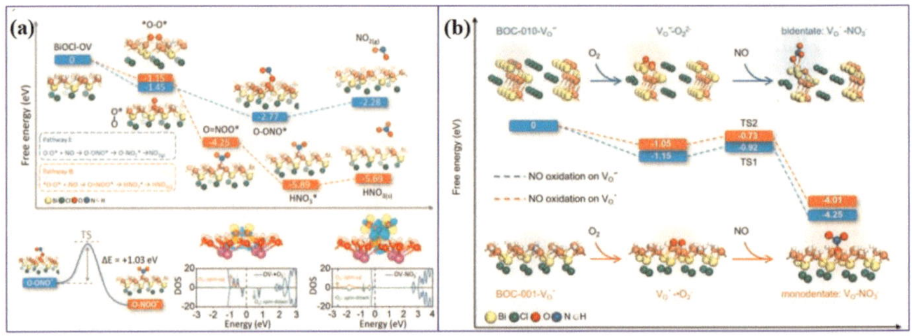

Figure 6.
(a) Free energy change against the reaction coordinate for the oxidation of NO by $\cdot O_2^-$ on BiOCl (001) surface in different geometries; (b) free energy change during O_2^{2-}- and $\cdot O_2^-$-mediated NO oxidation on the Vo of BOC-010 and BOC-001, respectively. Reproduced with permission from Refs. [9, 17], Copyright © 2018 and 2019, American Chemical Society.

staggered band alignment from the surface to the bulk and simultaneously generating more surface Vo. They proposed that O_2^- radicals enable the complete oxidation of NO into nitrate with high selectivity under visible-light irradiation (**Figure** 7) [40].

Zhang et al. reported highly efficient NO conversion over BiOI, and they found that NO removal pathways could be changed from nonselective oxidation to selective oxidation to produce nitrogen dioxide [41]. The —OH functionalization could enhance the reactants' activation capacity to exhibit the excellent photocatalytic NO conversion performances of BiOI to generate stable final products by activating O_2 molecules to generate active species [42]. For the heterojunction system, Lee et al. reported a visible-light heterojunction formed between insulator $SrCO_3$ and photosensitizer BiOI for NO conversion and proposed corresponding reaction pathways as $NO \rightarrow NO^+$ and $NO_2^+ \rightarrow$ nitrate or nitrite routes by *in situ* FTIR study [43]. To disclose the specific atomic interfacial electronic structure of heterostructure and its effect on the reaction, Sun et al. designed an insulator-based heterojunction $CaSO_4$-BiOI and further investigated the visible light NO conversions. Results suggest that the electronic environment of the interface/surface exerted great impacts on the active species formation, and the `intermediate transformation was supported by *in situ* DRIFT [44].

3.2.3 Bi-metal-based materials

The deposition of metals with local surface plasmon resonance effects such as Ag, Au, and Bi and Bi-vacancy on semiconductor surfaces is a common application, often in concert with Vo to promote photocatalytic NO oxidation (**Figure 8**) [45, 46]. Bismuth (Bi) is a commonly accepted semimetal that exhibits a highly anisotropic Fermi surface, low carrier density, small carrier effective mass, and long carrier mean free path, and investigated in the field of environmental pollution control widely due to its instinct characteristics such as broaden the optical ranges and promote the carrier separation. Although most of the research works are mainly focused on the NO

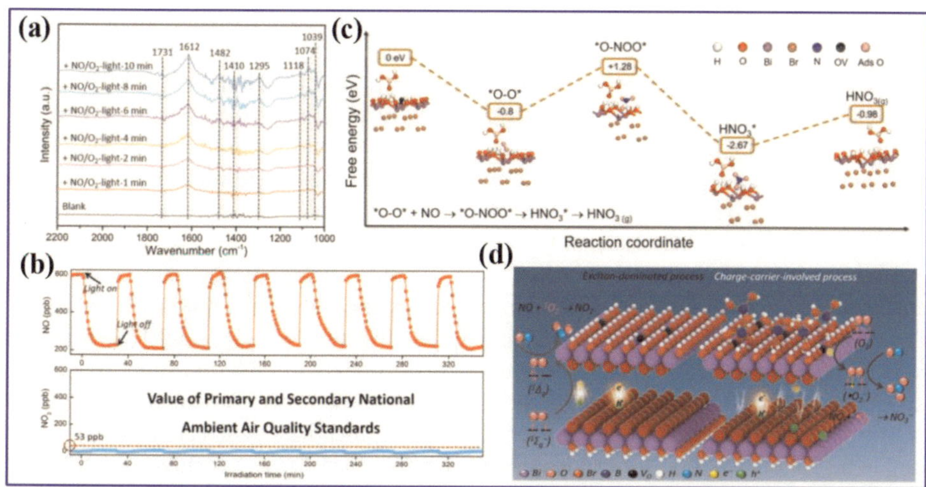

Figure 7.
(a) In situ *FTIR spectra of B-BiOBr for photocatalytic NO oxidation; (b) long-term NO removal; (c) free energy changes against the reaction coordinate for NO oxidation on modeled BiOBr and B-BiOBr surfaces, and (d) schematic illustration of the exciton-dominated and charge-carrier-involved photocatalytic NO oxidation processes. Reproduced with permission from Ref. [40], Copyright © 2022, American Chemical Society.*

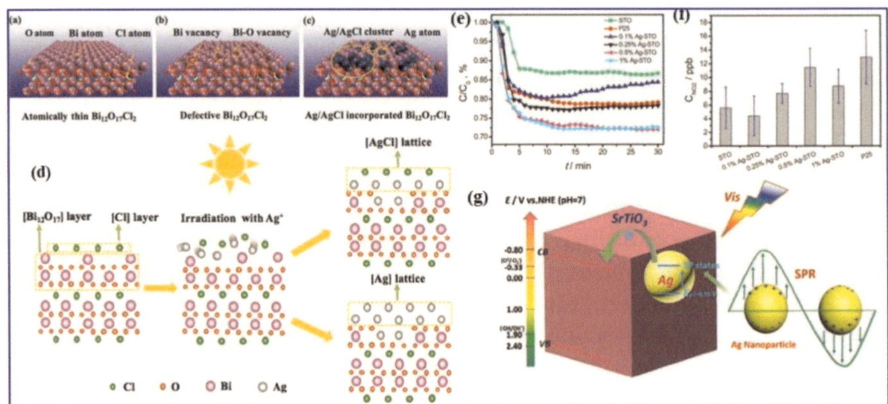

Figure 8.
(a–d) Schematic illustrations of the subnanometer Ag/AgCl clusters incorporated on atomically thin defective $Bi_{12}O_{17}Cl_2$ nanosheets via *rebinding with unsaturated Cl atoms; (e–g) visible light NO removal activity and proposed mechanism over Ag-$SrTiO_3$. Reproduced with permission from Refs. [45, 46], Copyright © 2016, 2021, American Chemical Society.*

removal efficiency instead of paying attention to the toxic byproduct formations during the NO conversion process. Many efforts have been devoted to tackling such a challenge. The defective Bi@$Bi_2Ti_2O_7$ photocatalyst was investigated for NO removal due to the co-effect of Bi-/Vo and displayed superior oxidation to NO_3^- compared to the defect-free $Bi_2Ti_2O_7$ counterpart [47]. Defective Bi/BiOBr nanoflowers were synthesized and further displayed 63% NO removal due to the effective carrier separation induced by bismuth and Vo. Importantly, the NO removal efficiency was negligibly affected by humidity, in which the generation of toxic NO_2 intermediate was reduced progressively from 87 to 29 ppb as the humidity increased from 5 to 100%, further indicating the significance of high humidity in promoting the transformation of toxic intermediate NO_2 to NO_3^- [48]. Due to the energetic hot electrons from the surface plasmon resonances of metallic Bi and superoxide generation from the molecular oxygen activation, Bi-metal-$BiPO_4$ (Bi-BPO) nanocomposites displayed 32.8% NO oxidation under illumination with visible light [49]. Dong et al. developed ultrathin Bi_2MoO_6 nanosheets modified by MoO_3 clusters, in which the ultrathin structure shortens the carrier transmission distance to reduce carrier recombination, while surface clusters highly favor the interfacial charge transfer, leading to fire-new electron migration pathway, and displayed highly efficient NO removal under relative humidity from 25 to 100% [50].

All these state-of-the-art indicate the significant roles of surface defects in improving NO conversion *via* facilitating the activation and adsorption of NO molecules on the surface of the photocatalysts. However, the instability and deactivation of surface Vo of photocatalysts in the continuous photocatalytic NO removal reaction results in a decrease in reaction selectivity, which needs to be improved with surface modifications. Taking advantage of surface defects and plasma effects of Bi metal, Dong et al. reported, the Bi nanoparticle decorated $Bi_2O_2CO_3$ nanosheets with Vo and obtained significantly enhanced NO conversion to generate NO_3^- with remarkably inhibited toxic intermediates NO_2 under visible-light illumination. The *in situ* DRFTS and DFT simulations indicate that the Bi nanoparticles and surface Vo act as active sites to activate the surface adsorbed O_2 and H_2O to produce molecular oxygen activation, and subsequently favored the NO oxidation to NO_3^-. and surface defects in promoting carrier separations are systematically investigated [51]. Bi-metal @ $Bi_2O_2[BO_2(OH)]$ with Vo was investigated for photocatalytic oxidation of NO under visible light and displayed the unique electron transfer covalent loop ($[Bi_2O_2]^{2+} \rightarrow$ Bi-metal $\rightarrow O_2^-$), which was confirmed by experimental and theoretical simulations. The Vo improve the charge separation efficiency and the yield of active oxygen species, while Bi-metal has functioned as electron donors to activate NO molecules and form NO^- and induces a new reaction path of NO $\rightarrow NO^- \rightarrow NO_3^-$ to achieve the harmless conversion of NO, effectively restraining the generation of noxious intermediates (NO_2, N_2O_4) [52]. A series of Bi and surface defects co-modified Zn_2SnO_4 were developed for NO removal and obtained enhanced NO conversions, which can be basically attributed to the synergistic effect of Vo and surface plasmon resonance (SPR) effects of Bi elements [53].

3.2.4 Other Bi-based materials

The roles of surface-interface investigated and displayed superior NO conversion in other Bi-based photocatalysts. For instance, Zhu et al. reported the Au nanoparticles loaded Vo-rich $Bi_4Ti_3O_{12}$ ($Au/Bi_4Ti_3O_{12}$) displayed 48% selectivity for NO removal and significantly reduced NO_2 production under visible light due to the obviously promoted carrier separation induced from the built-in electric field of surface plasmon resonance effect from Au nanoparticles [54]. The effects of Br^- on Vo construction over Bi_2MoO_6 with different facets exposed were systematically investigated, and the OVs concentration optimized Bi_2MoO_6 (BMO-001-Br) exhibited superior activity with 62.89% NO removal and 93.61% selectivity for complete NO oxidation [55]. The carbonate-intercalated defective Bi_2WO_6 facilitated the electron-hole pairs converting to reaction radical and reactants activation, including the H_2O oxidized to ·OH, O_2 reduced to $\cdot O_2^-$ and resulted in 55% oxidation of NO without generating secondary pollution of toxic intermediates [56]. To improve the transportation of the charge carriers, a Z-scheme heterojunction of 2D/2D BP/monolayer Bi_2WO_6 (MBWO) was designed and exhibited 67% NO oxidation owing to the intimate face-to-face contact between BP nanosheets and ultrathin MBWO [57]. Lee et al. reported the role of Vo in optimizing the performance of $Bi_2Sn_2O_{7-x}$ hollow nanocubes, which displayed 32% NO removal and suppression of NO_2 [58]. A Z-scheme n-$Bi_{12}SiO_{20}$/p-Bi_2S_3 displayed 56% photocatalytic NO removal and the diminished generation of NO_2 (3 ppb) within 4 min visible-light irradiation [59]. The n-p heterojunctions $Bi_{12}GeO_{20}$-Bi_2S_3 with surface Vo showed reinforced NO removal under visible light with 96% selectivity for NO_2^-/NO_3^- species, avoiding the generation of toxic NO_2 [60]. To tackle the bottlenecks of a few intrinsic active sites and inefficient carrier separation of photocatalysts during the NO removal, Lv et al. introduced Vo into Bi_3TaO_7 and achieved 5.4 folds higher NO removal efficiencies than bare Bi_3TaO_7. It was revealed that the intermediate products of defective-Bi_3TaO_7 are helpful to promote the deep oxidation of NO to NO_3^-, while pristine Bi_3TaO_7 is more likely to produce toxic intermediate NO_2, which greatly hinders the deep photocatalytic oxidation of NO [61]. Other Bi-Ti-based layered structured photocatalysts, such as $Pb_2Bi_4Ti_5O_{18}$ (**Figure 9**), $SrBi_2M_2O_9$ (M = Nb, Ta), and Bi-/$Bi_{12}TiO_{20}$, also displayed appreciable NO conversion and toxic intermediate generation, providing the encouraging results for the effective NO control in the field of environmental science [11, 62, 63].

All in all, Bi-based photocatalysts with pure and defective structures have been widely used for the removal of NO, during which the surface-interface controlling

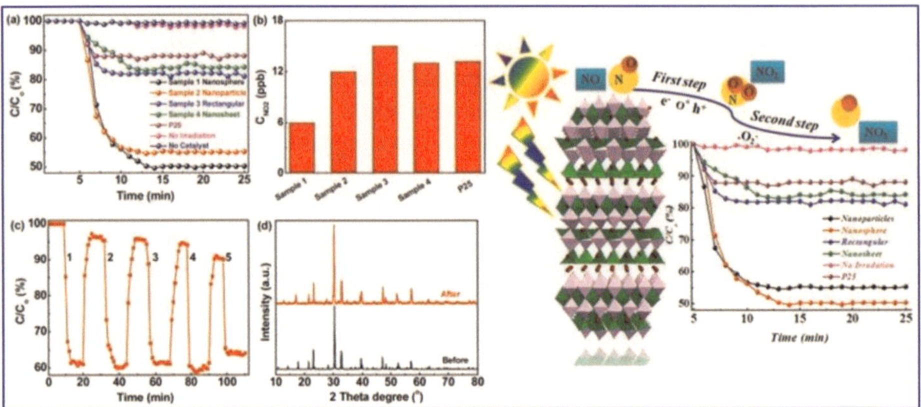

Figure 9.
Photocatalytic NO removal, the relative change in NO_2 concentration, photochemical stability, and NO removal mechanisms over layered Bi-based $Pb_2Bi_4Ti_5O_{18}$. Reproduced with permission from Ref. [61], Copyright © 2017, American Chemical Society.

such as surface defects, doing with plasmatic metal, or their coeffects for the removal of NO have been witnessed. The other semiconducting materials such as metal-free g-C_3N_4 and Ti-based perovskite have been extensively investigated, though reaction efficiency remains low due to massive and rapid recombination of photoinduced electron and holes, and cannot fulfill the requirements of both high efficiency and long-term robustness. Meanwhile, exact reaction mechanisms and conversion pathways of NO over such catalysts are still under debate, a lack of systematic investigations. More attention should be focused on the effective control of the toxic byproducts NO_2 and preventing catalyst deactivations.

4. Conclusions and perspectives

NO_X is harmful to both the environment and the human body, and thus the development of efficient NO_X control technology is of great significance for preventing and controlling air pollution. Compared with traditional physical or chemical adsorption technology, photocatalytic reactions can achieve the conversion or degradation of pollutants removal at room temperature and pressure with less secondary pollution, which is a new and efficient environment-friendly purification strategy. In view of the current state of photocatalytic NO removal, mainly for photooxidation, the lack of investigations on semiconductor-based photocatalysts and their industrial implementations on NO conversion is mainly due to the following factors:

1. To improve NO conversion efficiency, efforts should also be continued in other kinds of modification strategies such as vacancy and intercalation embedding engineering instead of merely being limited by the introduction of oxygen vacancies, Bi deposition, and the construction of Bi-based heterojunctions; moreover, although NO removal efficiency has been enhanced by the currently investigated semiconductors, their applications are still far from the industrial level; further investigations should be focused to achieving the greater degree of performance enhancement.

2. More attention still needs to be focused on developing nontoxic products that can oxidize NO to nitric acid or reduce it to N_2 to avoid the production of NO_2 to the maximum extent. Many issues of controlling the single reaction pathways on oxidation or reduction reaction, and its dependence on the surface structural and electronic features of the interface are still open. Thus, effective control of selectivity or NO conversion products and unearthing reaction mechanisms are highly desired.

3. The powdered photocatalytic semiconductors are not easily recoverable due to the occupation of surface-active sites by the final product NO_3^- as the reaction time increases, so NO_3^- should be removed by water washing and other means in a timely manner; Moreover, development of powder material solid-loading techniques, photocatalytic film macro-preparation, and preparation of photocatalytic devices are highly desired in anticipation of further adaptation of photocatalytic technology to practical applications

4. In the process of photocatalytic NO oxidation, $\cdot O_2^-$ is the most critical as the reactive oxygen species that can completely oxidize NO to NO_3^-, however, the role of $\cdot OH$ should not be wasted and neglected, and further oxidation of NO_2 from $\cdot OH$ activation is not a point worth exploring. In addition, the role of other ROS in this process and the mechanism are also worth discussing.

5. To study the reaction process and true mechanism of photocatalytic NO oxidation, more parameters continuous reactor, NO concentration, flow rate, temperature, water (relative humidity), catalysts loading, reactor sizes, air and light parameters (wavelength or intensity), detected conversion products and so on should be investigated deeply; meanwhile, more *in situ* characterizations and theoretical calculations at the atomic level should be served to disclose the real reaction pathways.

6. For the reduction pathway, the direct decomposition of NO into stoichiometric N_2 and O_2 under ambient conditions merely with light as the driving force remains offers one of the most ideal routes for NO treatment. However, its realization has been tremendously restricted by the huge activation barrier for both the cleavage of N=O bonds and the subsequent reconstruction of N≡N triple bonds, which has been hardly achieved under ambient conditions. More detailed investigations should be devoted to opening this gate.

7. During the photocatalytic NO removal, the chemical nature and concentration of particular matter, such as benzene, toluene, volatile organic compounds (VOCs), and other gases such as CO_2, SO_2, NH_3, and Hg^o, may compete with NO or reaction intermediates lead to reducing in the selectivity as well as removal efficiency of NO. To tackle this, future research needs to consider the interference of the above factors and other gases to establish more applicable methods. The possible solution is still closely related to the structure and interface characters of photocatalysts. Moreover, the source and concentration of interfering substances should also be considered to reduce the response of catalysts to interfering substances by adjusting the actual reaction conditions in advance.

Acknowledgements

Financial support by the National Natural Science Foundation of China (no. 22376008) is gratefully appreciated.

Conflict of interest

The authors declare that they have no known competing financial interests or personal relationships that could have appeared to influence the work reported in this contribution.

Appendices and nomenclature

WHO	World Health Organization
NO_x	NO nitrogen oxides
SCR	selective catalytic reduction
Photo-SCR	photo-selective catalytic reduction
H_2O_2	hydrogen peroxide
h^+	photoinduced holes
$^1O_2$	single line state oxygen
$\cdot O_2^-$	superoxide radicals
·OH	hydroxyl radicals
DRIFTS	*in situ* diffuse reflectance infrared Fourier transform spectroscopy
DFT	Density Functional Theory
Vo	oxygen vacancies or surface oxygen defects
2D/2D	two-dimensional/two-dimensional
EPR	electron paramagnetic resonance
SPR	surface plasmon resonance

Author details

Reshalaiti Hailili*, Zelong Li, Xu Lu and Xiaokaiti Reyimu*
MOE Key Laboratory of Enhanced Heat Transfer and Energy Conservation, Beijing Key Laboratory of Heat Transfer and Energy Conversion, Beijing University of Technology, Beijing, P. R. China

*Address all correspondence to: reshalaiti100@163.com; xiaokaiti100@163.com

References

[1] Max R. Data Review: How Many People Die from Air Pollution? World Health Organization (WHO); 2021. Available from: https://ourworldindata.org/data-review-air-pollution-deaths

[2] Han LP, Cai SX, Gao M, Hasegawa J-Y, Wang PL, et al. Selective catalytic reduction of NO_x with NH_3 by using novel catalysts: State of the art and prospects. Chemical Reviews. 2019;**119**:10916

[3] The Clean Air Act. Section 407, USA, Nitrogen oxides emission reduction program. 2004. Available from: https://www.govinfo.gov/content/pkg/USCODE-2011-title42/html/USCODE-2011-title42-chap85-subchapIV-A-sec7651f.htm

[4] Lyu M, Bao XF, et al. State-of-the-art outlook for light-duty vehicle emission control standards and technologies in China. Clean Technology Environment. 2020;**22**:757

[5] Air Quality Guidelines for Europe. World Health Organization Regional Office for Europe. Copenhagen: WHO; 2000

[6] He GZ et al. Polymeric vanadyl species determine the low-temperature activity of V-based catalysts for the SCR of NO_x with NH_3. Science Advances. 2018;**4**:eaau4637

[7] Chen S, Feng WJ, et al. A new strategy for plasma-catalytic reduction of NO to N_2 on the surface of modified Bi_2MoO_6. Chemical Engineering Journal. 2022;**440**:135754

[8] Wu QP, Van de Krol R. Selective photoreduction of nitric oxide to nitrogen by nanostructured TiO_2 photocatalysts: Role of oxygen vacancies and iron dopant. Journal of the American Chemical Society. 2012;**134**:9369

[9] Li H, Shang H, et al. Interfacial charging decharging strategy to efficient and selective aerobic NO oxidation on oxygen vacancy. Environmental Science & Technology. 2019;**53**:6964

[10] Li H, Zhu HJ, et al. Vacancy-rich and porous NiFe-layered double hydroxide ultrathin nanosheets for efficient photocatalytic NO oxidation and storage. Environmental Science & Technology. 2022;**56**:1771

[11] Hailili R, Wang Z-Q, Ji HW, et al. Mechanistic insights into the photocatalytic reduction of nitric oxide to nitrogen on oxygen-deficient quasi-two-dimensional bismuth-based perovskite. Environmental Science: Nano. 2022;**9**:1453

[12] Lim JB, Cha SH, Hong SB. Direct N_2O decomposition over Iron-substituted small-pore zeolites with different pore topologies. Applied Catalysis. B, Environmental. 2019;**243**:750

[13] Ford PC, Lorkovic IM. Mechanistic aspects of the reactions of nitric oxide with transition-metal complexes. Chemical Reviews. 2002;**102**:993

[14] Anpo M, Zhang SG, Mishima H, et al. Design of photocatalysts encapsulated within the zeolite framework and cavities for the decomposition of NO into N_2 and O_2 at normal temperature. Catalysis Today. 1997;**39**:159

[15] Hailili R, Ji HW, Wang KW, Dong XA, et al. ZnO with controllable oxygen vacancies for photocatalytic

nitrogen oxide removal. ACS Catalysis. 2022;**12**:10004

[16] Shang H, Li MQ, et al. Oxygen vacancies promoted the selective photocatalytic removal of NO with blue TiO_2 via simultaneous molecular oxygen activation and photogenerated hole annihilation. Environmental Science & Technology. 2019;**53**:6444

[17] Li H et al. Oxygen vacancies mediated complete visible light NO oxidation via side-on bridging superoxide radicals. Environmental Science & Technology. 2018;**52**:8659

[18] Dalton JS, Janes PA, et al. Photocatalytic oxidation of NO_x gases using TiO_2: A surface spectroscopic approach. Environmental Pollution. 2022;**120**:415

[19] Hunger M, Hüsken G, et al. Photocatalytic degradation of air pollutants: From modeling to large scale application. Cement and Concrete Research. 2010;**40**:313

[20] Wu QP et al. A dopant-mediated recombination mechanism in Fe-doped TiO_2 nanoparticles for the photocatalytic decomposition of nitric oxide. Catalysis Today. 2014;**225**:96

[21] Wang H, Li KL, et al. Photochemical transformation pathways of nitrates from photocatalytic NOx oxidation: Implications for controlling secondary pollutants. Environmental Science & Technology Letters. 2021;**8**:873

[22] Shang H, Chen ZY, et al. SO_2-enhanced nitrate photolysis on TiO_2 minerals: A vital role of photochemically reactive holes. Applied Catalysis. B, Environmental. 2022;**308**:121217

[23] Lei B, Cui W, Sheng JP, et al. Synergistic effects of crystal structure and oxygen vacancy on Bi_2O_3 polymorphs: Intermediates activation, photocatalytic reaction efficiency, and conversion pathway. Scientific Bulletin. 2020;**65**:467

[24] Hailili R, Reyimu X, Li ZL, et al. Tuning the microstructures of ZnO to enhance photocatalytic NO removal performances. ACS Applied Materials & Interfaces. 2023;**15**:23185

[25] Ma H, He Y, et al. Doping and facet effects synergistically mediated interfacial reaction mechanism and selectivity in photocatalytic NO abatement. Journal of Colloid and Interface Science. 2021;**604**:624

[26] Zhang J, Cui W, et al. B doped $Bi_2O_2CO_3$ hierarchical microspheres: Enhanced photocatalytic performance and reaction mechanism for NO removal. Catalysis Today. 2021;**380**:230

[27] Yuan CW, Chen RM, et al. La-doping induced localized excess electrons on $Bi_2O_2CO_3$ for efficient photocatalytic NO removal and toxic intermediates suppression. Journal of Hazardous Materials. 2020;**400**:123174

[28] Wang Y et al. Boosted photocatalytic efficiency of GQDs sensitized $Bi_2O_2CO_3$/β-Bi_2O_3 heterojunction via enhanced interfacial charge transfer. Chinese Chemical Letters. 2023;**34**:107967

[29] Rao F, Zhu GQ, et al. Maximizing the formation of reactive oxygen species for deep oxidation of NO via manipulating the oxygen-vacancy defect position on $Bi_2O_2CO_3$. ACS Catalysis. 2021;**11**:7735

[30] Huang Y, Zhu DD, et al. Synthesis of a $Bi_2O_2CO_3$/$ZnFe_2O_4$ heterojunction with enhanced photocatalytic activity for visible light irradiation-induced NO

removal. Applied Catalysis. B, Environmental. 2018;**234**:70

[31] Lu YF et al. Oxygen vacancy engineering of $Bi_2O_3/Bi_2O_2CO_3$ heterojunctions: Implications of the interfacial charge transfer, NO adsorption and removal. Applied Catalysis. B, Environmental. 2018; **231**:357

[32] Huo W, Cao T, et al. Facile construction of $Bi_2Mo_3O_{12}@Bi_2O_2CO_3$ heterojunctions for enhanced photocatalytic efficiency toward NO removal and study of the conversion process. Chinese. Journal of Catalysis. 2020;**41**:268

[33] Tang X, Huang ZA, et al. Mo promotes interfacial interaction and induces oxygen vacancies in 2D/2D of Mo-g-C_3N_4 and $Bi_2O_2CO_3$ photocatalyst for enhanced NO oxidation. Industrial and Engineering Chemistry Research. 2020;**59**:9509

[34] Liao JZ, Li KL, et al. Oxygen vacancies on the BiOCl surface promoted photocatalytic complete NO oxidation via superoxide radicals. Chinese Chemical Letters. 2020;**31**:2737

[35] Ren Q, He Y, et al. Photo-switchable oxygen vacancy as the dynamic active site in the photocatalytic NO oxidation reaction. ACS Catalysis. 2022;**12**:14015

[36] Shen T, Shi XK, et al. Photocatalytic removal of NO by light driven Mn_3O_4/BiOCl heterojunction photocatalyst: Optimization and mechanism. Chemical Engineering Journal. 2021;**408**:128014

[37] Geng Q, Xie HT, et al. Optimizing the electronic structure of BiOBr nanosheets via combined Ba doping and oxygen vacancies for promoted photocatalysis. Journal of Physical Chemistry C. 2021;**125**:8597

[38] Huo WC, Xu WN, et al. Motivated surface reaction thermodynamics on the bismuth oxyhalides with lattice strain for enhanced photocatalytic NO oxidation. Applied Catalysis. B, Environmental. 2021;**284**:119694

[39] Wu HZ, Yuan CW, et al. Mechanisms of interfacial charge transfer and photocatalytic NO oxidation on BiOBr/SnO_2 p-n heterojunctions. ACS Applied Materials & Interfaces. 2020;**12**: 43741

[40] Shi YB, Yang ZP, et al. Surface boronizing can weaken the excitonic effects of BiOBr nanosheets for efficient O_2 activation and selective NO oxidation under visible light irradiation. Environmental Science & Technology. 2022;**56**:14478

[41] Dong GH et al. Photocatalytic NO removal on BiOI surface: The change from nonselective oxidation to selective oxidation. Applied Catalysis. B, Environmental. 2015;**168**:490

[42] Yang WP, Ren Q, et al. Promotion mechanism of -OH group intercalation for NO_x purification on BiOI photocatalyst. Nanoscale. 2021;**13**:20601

[43] Wang H, Sun YJ, et al. Unraveling the mechanisms of visible light photocatalytic NO purification on earth-abundant insulator-based core-shell heterojunctions. Environmental Science & Technology. 2018;**52**:1479

[44] Wang H, Cui W, et al. Interfacial activation of reactants and intermediates on $CaSO_4$ insulator-based heterostructure for efficient photocatalytic NO removal. Chemical Engineering Journal. 2020;**390**:124609

[45] Zhu LL, Wu YF, et al. Tuning the active sites of atomically thin defective $Bi_{12}O_{17}C_{12}$ via incorporation of

subnanometer clusters. ACS Applied Materials & Interfaces. 2021;**13**: 9216-9223

[46] Zhang Q, Huang Y, et al. Visible-light-active plasmonic Ag-$SrTiO_3$ nanocomposites for the degradation of NO in air with high selectivity. ACS Applied Materials & Interfaces. 2016;**8**: 4165

[47] Zhu QH, Hailili R, et al. Efficient full spectrum responsive photocatalytic NO conversion at $Bi_2Ti_2O_7$: Co-effect of plasmonic Bi and oxygen vacancies. Applied Catalysis. B, Environmental. 2022;**319**:121888

[48] Xin Y, Zhu QH, et al. Photocatalytic NO removal over defective Bi/BiOBr nanoflowers: The inhibition of toxic NO_2 intermediate via high humidity. Applied Catalysis. B, Environmental. 2023;**324**: 122238

[49] Chen MJ et al. Synthesis and characterization of Bi-$BiPO_4$ nanocomposites as plasmonic photocatalysts for oxidative NO removal. Applied Surface Science. 2020; **513**:145775

[50] Ma H, He Y, et al. In situ loading of MoO_3 clusters on ultrathin Bi_2MoO_6 nanosheets for synergistically enhanced photocatalytic NO abatement. Applied Catalysis. B, Environmental. 2021;**292**: 120159

[51] Chen P, Liu HJ, et al. Bi-metal prevents the deactivation of oxygen vacancies in $Bi_2O_2CO_3$ for stable and efficient photocatalytic NO abatement. Applied Catalysis. B, Environmental. 2020;**264**:118545

[52] Zhou X et al. Efficient NO removal and photocatalysis mechanism over Bi-metal @$Bi_2O_2[BO_2(OH)]$ with oxygen vacancies. Journal of Hazardous Materials. 2022;**436**:129271

[53] Gu ML, Li YH, et al. Bismuth nanoparticles and oxygen vacancies synergistically attired Zn_2SnO_4 with optimized visible-light-active performance. Nano Energy. 2021;**80**: 105415

[54] Liu ZX, An YR, et al. Au nanoparticles modified oxygen-vacancies-rich $Bi_4Ti_3O_{12}$ heterojunction for efficient photocatalytic NO removal with high selectivity. Journal of Alloy Compounds. 2023;**942**:169018

[55] Wang SY, Ding X, et al. Insight into the effect of bromine on facet-dependent surface oxygen vacancies construction and stabilization of Bi_2MoO_6 for efficient photocatalytic NO removal. Applied Catalysis. B, Environmental. 2020;**265**: 118585

[56] Huo WC et al. Carbonate-intercalated defective bismuth tungstate for efficient photocatalytic NO removal and promotion mechanism study. Applied Catalysis. B, Environmental. 2019;**254**:206

[57] Hu JD et al. Z-scheme 2D/2D heterojunction of black phosphorus/ monolayer Bi_2WO_6 nanosheets with enhanced photocatalytic activities. Angewandte Chemie International Edition. 2019;**58**:2073

[58] Lu YF et al. Oxygen vacancy-dependent photocatalytic activity of well-defined $Bi_2Sn_2O_{7-x}$ hollow nanocubes for NO_x removal. Environmental Science. Nano. 2021, 1927:8

[59] Chang F et al. Mechanical ball-milling preparation and superior photocatalytic NO elimination of Z-scheme $Bi_{12}SiO_{20}$-based

heterojunctions with surface oxygen vacancies. Journal of Cleaner Production. 2022;**380**:135167

[60] Chang F, Wang XM, et al. Enhanced photocatalytic NO removal with the superior selectivity for NO_2^-/NO_3^- species of $Bi_{12}GeO_{20}$-based composites via a ball-milling treatment: Synergetic effect of surface oxygen vacancies and n-p heterojunctions. Composites Part B Engineering. 2022;**231**:109600

[61] Liu L, Ouyang P, et al. Insight into the mechanism of deep NO photo-oxidation by bismuth tantalate with oxygen vacancies. Journal of Hazardous Materials. 2022;**439**:129637

[62] Hailili R, Dong GH, et al. Layered perovskite $Pb_2Bi_4Ti_5O_{18}$ for excellent visible light-driven photocatalytic NO removal. Industrial and Engineering Chemistry Research. 2017;**56**:2908

[63] Li N et al. Modulation of photocatalytic activity of $SrBi_2Ta_2O_9$ nanosheets in NO removal by tuning facets exposure. Journal of Materials Science and Technology. 2022;**122**:91

Chapter 3

Research on Partial Nitritation and Anaerobic Ammonium Oxidation Process

Wenqiang Wang, Dong Li, Shuai Li, Huiping Zeng and Jie Zhang

Abstract

In recent years, the partial nitritation and anaerobic ammonium oxidation (PN/A) process has been widely appreciated by many countries around the world. As an autotrophic nitrogen removal process, this process can save more than 60% of the aeration energy consumption, reduce 80% of the residual sludge yield, and do not need to add additional carbon sources. However, this process is faced with several kinds of problems. This paper summarizes several effects of operating parameters on the inhibition of NOB in municipal wastewater treatment, implications of the reactor configuration and operation, and fixed film processes vs. suspended growth systems. The fixed film processes based on Anammox granular sludge and AOB flocculent sludge are alternative. Finally, a new strategy of continuous flow PN/A process with partial nitrification flocculent sludge and Anammox granular sludge was proposed.

Keywords: Anammox, wastewater treatment, PN/A, granular sludge, mixed reactor

1. Introduction

Anaerobic ammonia oxidation (Anammox) bacteria play a key role in the Earth's nitrogen cycle [1]. Compared with Anammox, the traditional biological denitrification technology widely used today has some shortcomings, such as high energy consumption, the need for additional organic carbon sources, and the inability to control the production of greenhouse gases [2]. Therefore, compared with the traditional nitrification and denitrification systems, Anammox technology has the advantages of no aeration, extra organic carbon source, and low surplus sludge yield, making Anammox a research hotspot for environmental protection. In the application of actual municipal wastewater [3], it is found that the lack of nitrite electron acceptor in actual wastewater is the main bottleneck of the application of Anammox in mainstream wastewater treatment. In addition to the artificial addition of nitrite, most of the current Anammox coupling processes focus on partial nitritation and anaerobic ammonium oxidation (PN/A) [4]. In practical application, partial nitritation (NH_4^+-N → NO_2^--N) can provide NO_2^--N for Anammox, thus forming a PN/A process. First, about 55% of the ammonia nitrogen in the wastewater is oxidized to nitrite under the action of ammonia oxidizing bacteria (AOB), and then the generated nitrite and the remaining ammonia nitrogen generate nitrogen under the action of anaerobic ammonia oxidation bacteria (AnAOB), to achieve the removal of TN. The total reaction equation [5] is:

$$NH_4^+ + 0.85O_2 \rightarrow 0.11NO_3^- + 0.44N_2 + 0.14H^+ + 1.43H_2O \quad (1)$$

In recent years, PN/A process has been widely appreciated by many countries around the world. As an autotrophic nitrogen removal process, this process can save more than 60% of the aeration energy consumption, reduce 80% of the residual sludge yield, and does not need to add additional carbon sources [6, 7]. However, this process is faced with the problems of long age and insufficient retention capacity of AnAOB sludge, competition of AOB from NOB, and high C/N ratio leading to massive reproduction of HB in sludge.

This paper summarizes several effects of operating parameters on the inhibition of NOB in municipal wastewater treatment, implications of the reactor configuration and operation, and fixed film processes vs. suspended growth systems. Finally, a new strategy of continuous flow PN/A process with partial nitrification flocculent sludge and Anammox granular sludge was proposed.

2. Effects of operating parameters on the inhibition of NOB in municipal wastewater treatment

2.1 Dissolved oxygen (DO)

Continuous aeration mode keeps DO at a low level in the CANON process reactor. However, Liu et al. found through experiments that under long-term low DO (0.16 ~ 0.37 mg/L) operation, the dissolved oxygen affinity coefficient Ko_2, AOB will be higher than Ko_2, NOB, that is, NOB is more competitive than AOB. As a result, part of NO_2^--N in the reactor is further converted to NO_3^--N, and the lack of stable nitrite supply greatly endangers the overall nitrogen removal performance of autotrophic nitrogen removal process [8]. In view of this phenomenon, Regmi et al. believed that Nitrobacter would increase significantly more than Nitrospira in a long-term state of low DO [9]. Nitrospira is the strategy-K (low specific proliferation rate, high matrix affinity), while Nitrobacter is the strategy-R (high specific proliferation rate, low matrix affinity). As a result, Nitrospira is more amenable to compete with DO. The previous thought that Ko_2, NOB was higher than Ko_2, AOB might be due to the dominance of Nitrobacter. Controlling DO concentrations of medium (<1.0 mg/L) and low (<0.5 mg/L) is indeed beneficial to Nitrosomonas, but it inhibits the Nitrobacter of strategy-R rather than the Nitrospira of strategy-K (which grows at a rate close to the maximum).

2.2 Transient hypoxia

Transient hypoxia caused by on/off aeration has been proven to be an effective method to inhibit NOB. The lag time of NOB activity at the beginning of aeration section is due to the following two reasons: (1) Lack of one or two substrates (nitrite and oxygen) [10]. (2) In the aerobic environment after a brief period of hypoxia, compared with AOB, NOB faces metabolic mechanism inactivation and adaptive lag in the recovery period [11]. The delay of nitro Spirillum activity after anoxic period (5–15 min) effectively inhibited NOB in PN/A process based on SBR [11]. Intermittent aeration is also effective in integrated fixed-film activated sludge (IFAS) process [12]. However, intermittent aeration is not conducive to the stability of effluent quality in continuous flow process and frequent opening and closing of blower will increase the failure rate of equipment. However, in a continuous system that consumes nitrous oxide in time, an aerobic/aerobic alternating strategy can effectively carry out partial nitrification [11, 13].

2.3 Starvation process

Studies have shown that nitrite accumulation occurs when nitrification system is restarted after being idle for a period of time. The attenuation rate of AOB is smaller than that of NOB in the starvation process [14, 15]. Jia et al. [16] found that AOB has a unique hunger response strategy, and its cells are in a state of readiness at any time. Once the substrate appears, it can produce substrate invertase (gene transcription speed is fast), so that AOB can quickly recover its activity from starvation. However, NOB does not have this ability, so AOB has a stronger ability to adapt to environmental changes than NOB.

2.4 Aerobic sludge residence time

In the side-flow anaerobic process, the growth rate of AOB is higher than that of NOB, so short-time aerobic SRT is used to inhibit and flush NOB [17]. In the activated sludge process with temperatures between 28 and 30°C, aerobic SRT of 2.5 d is one of the main factors to ensure stable nitrogen removal in the Singapore Changi Regenerating water plant [18]. The control aerobic SRT of large-scale activated sludge denitrification process in the Southwest Wastewater Treatment Plant in St. Petersburg was 3.5 d, which is another reference case of this operation mode [19]. In the mainstream process, Blackburne and Oleszkiewicz found that shortening SRT was beneficial for AOB growth [19, 20]. Stinson et al. used short-term aerobic SRT as an intervention to achieve inhibition of NOBs at moderate and low temperatures [21].

2.5 Real-time aeration control

A variety of real-time aeration control strategies have been developed and applied to suppress NOBs by controlling DO or aeration volume. Response parameters related to DO or oxygen supply include ammonia nitrogen flux and dpH/dt [22]. In addition, frequent opening and closing of mechanical equipment such as blowers or pumps can increase the failure rate and cause serious operational problems, indicating that the reliability of key equipment is very important to A stable mainstream PN/A process.

3. Implications of reactor configuration and operation

3.1 Carbon pretreatment process

In view of the high carbon-nitrogen ratio of municipal sewage, carbon pretreatment is usually introduced. At present, there are three carbon pretreatment processes at home and abroad: (1) High-rate activated sludge (HRAS). For example, in the Strass sewage treatment plant, A stage of activated sludge process (SRT $\approx$ 0.5d,

HRT ≈ 0.5 h) can guarantee the removal rate of COD of 60% [23]. (2) Chemical-intensive pretreatment can remove about 80% ~ 90% TSS and 50% ~ 70% COD; (3) Methanogenic fermentation pretreatment, designed to maximize energy recovery in UASB reactors [24]. Generally, the three reasonably designed pretreatment processes can satisfy the PN/A influent COD/N ≤ 2 ~ 3.

3.2 One-stage and two-stage processes

The one-stage PN/A process performs PN/A in one reactor, while the two-stage PN/A process separates the PN/A reaction in two reactors. The combination of PN and Anammox reactions in a reactor significantly reduces infrastructure and operating costs compared to a two-stage process [25]. One-stage reactors tend to operate under nitrite restriction and low DO concentrations.

One-stage PN/A process is also known as CANON (Completely autotrophic nitrogen removal over nitrite, CANON) granular sludge has a regular shape and compact, dense structure, high sludge concentration, good settling performance, etc. Winkler et al. believed that in typical CANON granular sludge, AOB bacteria were usually distributed in the outer layer of particles permeable by dissolved oxygen, while AnAOB bacteria were distributed in the inner anaerobic zone of particles [23]. However, with the in-depth study of CANON granular sludge, some scholars found that AOB and AnAOB bacteria co-existed in CANON granular sludge, without specific spatial distribution rules, the two bacteria interleave each other, and the nitrite matrix produced by AOB ammonia-nitrogen does not need to be transferred through a long chain, and is degraded as the substrate of AnAOB bacteria in a short time. The whole autotrophic nitrogen removal process can be efficiently completed [26, 27]. Chen's research shows that the combination of AnAOB and AOB forms a special olivar-shaped structure: (1) AnAOB mainly gathers inside the particle to form the kernel of the particle, AOB forms a thick wall in the outer layer of the particle, (2) AnAOB gathers into multiple clusters, and AOB is relatively evenly distributed in the whole particle without any clusters. (3) The many cracks clearly observed in the Anammox particles are likely conduits through which substrates and wastes flow. In this special structure, part of influent NH_4^+-N is oxidized to NO_2^--N by AOB in the particle surface layer, and Anammox in the particle core uses residual NH_4^+-N and generated NO_2^--N. In addition, the consumption of O_2 by AOB covered by the outer layer of particles provides protection for AnAOB from inhibition of O_2 and other environmental factors [27]. Statistics show that more than 50% of PN/A reactors operate in SBR mode, 88% of wastewater plants operate using single-stage systems, and 75% are used to treat sideflow municipal wastewater. Solid film substrate transport, aeration control, and nitrate generation are major operational difficulties [28]. Moreover, the single-stage CANON is mostly used in side-flow processes.

4. Fixed film processes vs. suspended growth systems

4.1 Suspended sludge reactor

The Changi Reclaimed Water Plant in Singapore and the Strass Sewage Treatment Plant in Austria are two existing PN/A processes using suspended sludge reactors. In both cases, AnAOB is not affected by oxygen in the anoxic zone, and heterotrophic denitrification makes an important contribution to nitrogen removal at high C/N influent ratios. One of the most common problems of flocculent sludge is poor settling [29]. Small floc cannot settle and separate well, resulting in increased turbidity of effluent [30].

4.2 Biofilm reactor

Biofilm reactor includes biological drip filter, fixed membrane biofilm reactor, fluidized bed reactor, biological rotary table, and moving bed biofilm reactor. The typical problem with biofilms is mass transfer restriction, in which only the bacteria in the outer layer of the biofilm can promote substrate removal [31]. The biofilm mainly grows on the surface of the filler, and the filler is prone to clogging during the actual operation, which is not conducive to the long-term operation of the process [32]. In addition, the cost of biofilm carriers is also a limitation of their application in large-scale wastewater treatment plants [33].

4.3 Granular sludge reactor

Bacterial aggregation in granular sludge reactors increased biomass concentration [34], as well as biomass retention [35] and tolerance to environmental stress [36]. Granular sludge can fix more AnAOB population under aeration conditions [37, 38]. Compared with floc and biofilm, particles have the advantages of dense structure and do not need to be attached to the surface of the carrier, so they are favored by the engineering field. However, the disadvantage of Anammox granules is that the low cell yield and growth rate of AnAOB lead to long sludge age, and HB grows excessively on autotrophic bacteria under long sludge age conditions.

4.4 Mixed reactor

The mixed reactor has a high concentration of suspended sludge in the liquid phase. The biomass of liquid and solid (biofilm or granular sludge) phases plays an important role in microbial transformation. Based on the theory that AnAOB mainly exists in particles or biofilms, while NOB and HB mainly exist in floc [39]. Compared with the biofilm reactor, AOB, NOB, and HB are mainly suspended in the liquid phase, while AnAOB mainly exists in the biofilm. It has been reported that 60% of aerobic reactions are achieved in the liquid phase, while AnAOB activity occurs almost exclusively in biofilms (>96.5%) [24]. The high suspended sludge concentration in the liquid phase of the mixed reactor significantly reduces the diffusion limit compared with the single particle or biofilm reactor, thus inhibiting the competition of AOB by controlling the low concentration level of DO in the liquid phase. In addition, the sludge age of suspended sludge can be controlled independently of particles or biofilms, which facilitates the washing of HB and NOB, and can tolerate higher influent COD/N [40].

It has been reported that the nitrogen removal capacity of IFAS process co-existing with flocculent sludge and biofilm is three to four times higher than that of MBBR process consisting of biofilm only [37, 41]. Compared with biofilms, particles have the advantage of dense structure and do not need to be attached to the surface of the medium. However, the process of coexisting Anammox particles with continuous flow flocculent sludge has not been reported. If the flocculent sludge with AOB as the main body and Anammox granular sludge is integrated into a continuous flow system, the nitrogen removal effect can be ensured, and the excess AOB and NOB short sludge age bacteria can be eliminated through the flocculent sludge to retain AnAOB, and the NOB can also be washed. Studies have shown that at low concentrations, intermittent anaerobic/aerobic = (15 ~ 20 min)/(5 ~ 15 min) alternating can effectively inhibit NOB [10]. Therefore, continuous flow anaerobic/aerobic alternating operation can inhibit NOB.

In addition to the difficulties of NOB suppression and panning, internal reflux and external reflux are often needed in continuous flow. In this process, the granular sludge of Anammox will inevitably be broken and disintegrated, which will also face the problem of AnAOB loss. If the problem of reflux is not solved, granular sludge can not be used effectively in the reactor with continuous push flow. At present, the methods of retaining granular sludge include membrane screening method and settling selection method, which represent the reactor membrane reactor and granular sludge selector respectively. Compared with membrane reactor, granular sludge selectors can pass flocculent sludge, not easy to clog, and is easy to maintain. If the granular sludge selector based on sedimentation selection is installed in the aerobic zone to retain Anammox particles and ensure the circulating flow of flocculent sludge, the reflux problem can be solved.

To sum up, in order to run the continuous autotrophic nitrogen removal process stably, Anammox granular sludge can be retained in the aerobic zone to participate in nitrite degradation but not reflux according to the anaerobic/aerobic process, so as to ensure the integrity of Anammox particles and effective retention of AnAOB while consuming most nitrite in time. The AOB in the floc is pumped back to alternate anaerobic/aerobic operation for nitrosation reaction. Of course, due to the mixed reactor, a small amount of AnAOB will exist in the floc, and the amount of AnAOB will not be too high because of the short floc sludge age. A small amount of AOB will also adhere to the Anammox particles to form a protective layer to consume dissolved oxygen.

5. Conclusions

This paper comprehensively describes the effects of operating parameters on the inhibition of NOB in municipal wastewater treatment, implications of reactor configuration and operation, and focuses on the discussion of fixed film processes versus suspended growth systems. A mixed urban sewage continuous flow PN/A process based on Anammox granular sludge is proposed. This process seems to have many advantages in theory, and it will be validated in experiments next.

Acknowledgements

This work was supported by China Postdoctoral Science Foundation (2022M720330) and Beijing Outstanding Young Scientist Program (BJJWZYJH01201910005019).

Conflict of interest

The authors declare no conflict of interest.

Author details

Wenqiang Wang[1*], Dong Li[1], Shuai Li[1], Huiping Zeng[1] and Jie Zhang[1,2]

1 Key Laboratory of Water Science and Water Environment Recovery Engineering, Beijing University of Technology, Beijing, China

2 State Key Laboratory of Urban Water Resource and Environment, Harbin Institute of Technology, Harbin, China

*Address all correspondence to: wangwenqiang@bjut.edu.cn

References

[1] Kuypers MMM, Sliekers AO, Lavik G, Schmid M, Jørgensen BB, Kuenen JG, et al. Anaerobic ammonium oxidation by anammox bacteria in the black sea. Nature. 2003;**422**:608-611. DOI: 10.1038/nature01472

[2] Kartal B, Kuenen JG, van Loosdrecht MCM. Sewage treatment with anammox. Science. 2010;**328**:702-703. DOI: 10.1126/science.1185941

[3] Yang Y, Azari M, Herbold CW, Li M, Chen H, Ding X, et al. Activities and metabolicversatility of distinct anammox bacteria in a full-scale waste water treatment system. Water Research. 2021;**206**:12. DOI: 10.1016/j.watres.2021.117763

[4] Agrawal S, Weissbrodt DG, Annavajhala M, Jensen MM, Arroyo JMC, Wells G, et al. Time to act-assessing variations in qPCR analyses in biological nitrogen removal with examples from partial nitritation/anammox systems. Water Research. 2021;**190**:8. DOI: 10.1016/j.watres.2020.116604

[5] Sliekers AO, Derwort N, Campos-Gomez JL, Strous M, Kuenen JG, Jetten M. Completely autotrophic nitrogen removal over nitrite in one single reactor. Water Research. 2002;**36**:2475-2482. https://doi.org/DOI. DOI: 10.1016/S0043-1354(01)00476-6

[6] Jetten MSM, Horn SJ, van Loosdrecht MCM. Towards a more sustainable municipal wastewater treatment system. Water Science and Technology. 1997;**35**:171-180. DOI: 10.1016/S0273-1223(97)00195-9

[7] Wett B. Development and implementation of a robust deammonification process. Water Science and Technology. 2007;**56**:81-88. DOI: 10.2166/wst.2007.611

[8] Liu G, Wang J. Long-term low do enriches and shifts nitrifier community in activated sludge. Environmental Science & Technology. 2013;**47**:5109-5117. DOI: 10.1021/es304647y

[9] Regmi P, Miller MW, Holgate B, Bunce R, Park H, Chandran K, et al. Control of aeration, aerobic SRT and COD input for mainstream nitritation/denitritation. Water Research. 2014;**57**:162-171. DOI: 10.1016/j.watres.2014.03.035

[10] Gilbert EM, Agrawal S, Brunner F, Schwartz T, Horn H, Lackner S. Response of different Nitrospira species to anoxic periods depends on operational do. Environmental Science & Technology. 2014;**48**:2934-2941. DOI: 10.1021/es404992g

[11] Kornaros M, Dokianakis SN, Lyberatos G. Partial nitrification/denitrification can be attributed to the slow response of nitrite oxidizing bacteria to periodic anoxic disturbances. Environmental Science & Technology. 2010;**44**:7245-7253. DOI: 10.1021/es100564j

[12] Trojanowicz K, Plaza E, Trela J. Pilot scale studies on nitritation-anammox process for mainstream wastewater at low temperature. Water Science and Technology. 1998;**37**:135-142. DOI: 10.2166/wst.2015.551

[13] Ge S, Peng Y, Qiu S, Zhu A, Ren N. Complete nitrogen removal from municipal wastewater via partial nitrification by appropriately alternating anoxic/aerobic conditions in a continuous plug-flow step feed

process. Water Research. 2014;**55**:95-105. DOI: 10.1016/j.watres.2014.01.058

[14] Feng C, Lotti T, Lin Y, Malpei F. Extracellular polymeric substances extraction and recovery from anammox granules: Evaluation of methods and protocol development. Chemical Engineering Journal. 2019;**374**:112-122. DOI: 10.1016/j.cej.2019.05.127

[15] Liu YQ, Liu Y, Tay JH. The effects of extracellular polymeric substances on the formation and stability of biogranules. Applied Microbiology and Biotechnology. 2004;**65**:143-148. DOI: 10.1007/s00253-004-1657-8

[16] Jia F, Yang Q, Liu X, Li X, Li B, Zhang L, et al. Stratification of extracellular polymeric substances (EPS) for aggregated anammox microorganisms. Environmental Science & Technology. 2017;**51**:3260-3268. DOI: 10.1021/acs.est.6b05761

[17] Hellinga CSAMJ. The SHARON process: An innovative method for nitrogen removal from ammonium rich waste water. Water Science and Technology. 1998;**37**:135-142

[18] Cao Y, Kwok BH, van Loosdrecht MCM, Daigger GT, Png HY, Long WY, et al. The occurrence of enhanced biological phosphorus removal in a 200,000 m3/day partial nitration and anammox activated sludge process at the changi water reclamation plant, Singapore. Water Science and Technology. 2017;**75**:741-751. DOI: 10.2166/wst.2016.565

[19] Blackburne RJ. Nitrifying Bacteria Characterisation to Identify and Implement Factors Leading to Nitrogen Removal Via Nitrite in Activated Sludge Processes. Australia: The University of Queensland; 2006

[20] Yuan Q, Oleszkiewicz JA. Low temperature biological phosphorus removal and partial nitrification in a pilot sequencing batch reactor system. Water Science and Technology. 2011;**63**:2802-2807. DOI: 10.2166/wst.2011.609

[21] Stinson BMSBC, Mokhyerie YDCH. Roadmap toward energy neutrality & chemical optimization at enhanced nutrient removal facilities. In: Proceedings of WEF/IWA Nutrient Removal and Recovery: Trends in Resource Recovery and Use. 28-31 July 2013. Vancouver, Canada: WEF; 2013

[22] Yang Q, Peng Y, Liu X, Zeng W, Mino T, Satoh H. Nitrogen removal via nitrite from municipal wastewater at low temperatures using real-time control to optimize nitrifying communities. Environmental Science & Technology. 2007;**41**:8159-8164. DOI: 10.1021/es070850f

[23] Winkler MKH, Kleerebezem R, van Loosdrecht MCM. Integration of anammox into the aerobic granular sludge process for main stream wastewater treatment at ambient temperatures. Water Research. 2012;**46**:136-144. DOI: 10.1016/j.watres.2011.10.034

[24] Malovanyy A, Yang J, Trela J, Plaza E. Combination of upflow anaerobic sludge blanket (UASB) reactor and partial nitritation/anammox moving bed biofilm reactor (MBBR) for municipal wastewater treatment. Bioresource Technology. 2015;**180**:144-153. DOI: 10.1016/j.biortech.2014.12.101

[25] Vlaeminck SE, De Clippeleir H, Verstraete W. Microbial resource management of one-stage partial nitritation/anammox. Microbial Biotechnology. 2012;**5**:433-448. DOI: 10.1111/j.1751-7915.2012.00341.x

[26] Chen R, Ji J, Chen Y, Takemura Y, Liu Y, Kubota K, et al. Successful operation performance and syntrophic micro-granule in partial nitritation and anammox reactor treating low-strength ammonia wastewater. Water Research. 2019;**155**:288-299. DOI: 10.1016/j.watres.2019.02.041

[27] Li X, Sung S. Development of the combined nitritation-anammox process in an upflow anaerobic sludge blanket (UASB) reactor with anammox granules. Chemical Engineering Journal. 2015;**281**:837-843. DOI: 10.1016/j.cej.2015.07.016

[28] Lackner S, Gilbert EM, Vlaeminck SE, Joss A, Horn H, van Loosdrecht MCM. Full-scale partial nitritation/anammox experiences–an application survey. Water Research. 2014;**55**:292-303. DOI: 10.1016/j.watres.2014.02.032

[29] Wang G, Xu X, Zhou L, Wang C, Yang F. A pilot-scale study on the start-up of partial nitrification-anammox process for anaerobic sludge digester liquor treatment. Bioresource Technology. 2017;**241**:181-189. DOI: 10.1016/j.biortech.2017.02.125

[30] Zhang G, Zhang P, Yang J, Chen Y. Ultrasonic reduction of excess sludge from the activated sludge system. Journal of Hazardous Materials. 2007;**145**:515-519. DOI: 10.1016/j.jhazmat.2007.01.133

[31] Henze M, Van Loosdrecht MC, Ekama GA, Brdjanovic D. Biological Wastewater Treatment. London, UK: IWA; 2008

[32] Zhang Z, Liu S, Miyoshi T, Matsuyama H, Ni J. Mitigated membrane fouling of anammox membrane bioreactor by microbiological immobilization. Bioresource Technology. 2016;**201**:312-318. DOI: 10.1016/j.biortech.2015.11.037

[33] Zhao Y, Liu D, Huang W, Yang Y, Ji M, Nghiem LD, et al. Insights into biofilm carriers for biological wastewater treatment processes: Current state-of-the-art, challenges, and opportunities. Bioresource Technology. 2019;**288**:121619. DOI: 10.1016/j.biortech.2019.121619

[34] Rittmann BE. Biofilms, active substrata, and me. Water Research. 2018;**132**:135-145. DOI: 10.1016/j.watres.2017.12.043

[35] Winkler MKH, Bassin JP, Kleerebezem R, van der Lans RGJM, van Loosdrecht MCM. Temperature and salt effects on settling velocity in granular sludge technology. Water Research. 2012;**46**:3897-3902. DOI: 10.1016/j.watres.2012.04.034

[36] Adav SS, Lee D, Show K, Tay J. Aerobic granular sludge: recent advances. Biotechnology Advances. 2008;**26**:411-423. DOI: 10.1016/j.biotechadv.2008.05.002

[37] Veuillet FZPSD, Ochoa JLR. Mainstream deammonification using ANITA™Mox process. In: Proceedings of IWA Specialist Conference Nutrient Removal and Recovery: Moving Innovation into Practice. 18-21 May 2015. Gdańsk, Poland: IWA; 2015

[38] Xu G, Zhou Y, Yang Q, Lee ZM, Gu J, Lay W, et al. The challenges of mainstream deammonification process for municipal used water treatment. Applied Microbiology and Biotechnology. 2015;**99**:2485-2490. DOI: 10.1007/s00253-015-6423-6

[39] Hubaux N, Wells G, Morgenroth E. Impact of coexistence of flocs and biofilm on performance of combined nitritation-anammox granular sludge reactors. Water Research. 2015;**68**:127-139. DOI: 10.1016/j.watres.2014.09.036

[40] Winkler MKH, Kleerebezem R, Kuenen JG, Yang J, van Loosdrecht MCM. Segregation of biomass in cyclic anaerobic/aerobic granular sludge allows the enrichment of anaerobic ammonium oxidizing bacteria at low temperatures. Environmental Science & Technology. 2011;**45**:7330-7337. DOI: 10.1021/es201388t

[41] Veuillet F, Lacroix S, Bausseron A, Gonidec E, Ochoa J, Christensson M, et al. Integrated fixed-film activated sludge anita™mox process–a new perspective for advanced nitrogen removal. Water Science and Technology. 2014;**69**:915-922. DOI: 10.2166/wst.2013.786

Chapter 4

The Contribution of Autotrophic Nitrogen Oxidizers to Global Nitrogen Conversion

Hui-Ping Chuang, Akiyoshi Ohashi and Hideki Harada

Abstract

The accumulation of ammonium (NH_4^+-N) and nitrous oxide (N_2O-N) in the environment is causing concern due to their ecological impacts and contribution to global warming. Autotrophic nitrogen oxidizers, including aerobic ammonium-oxidizing archaea and bacteria, anaerobic ammonium oxidizer and nitrite oxidizers, play a crucial role in the nitrogen cycle by facilitating the removal of nitrogenous residues from the environment. Nitrogen oxides (NOx) like nitrite (NO_2^--N) and nitrate (NO_3^--N) are produced as key immediate products during the conversion of NH_4^+-N or N_2O-N. Additionally, these autotrophic microbes utilize carbon dioxide (CO_2) for cell synthesis, thereby mitigating the greenhouse effect. Preliminary results pointed out that nitrogen oxidizers could effectively remove NH_4^+-N and NO_x from sewage and wastewater systems at the loading rate lower than 0.5 kg N/m^3-day. Moreover, this family could also reduce the greenhouse N_2O-N through oxidizing pathway, attaining the maximum reduction of 25.2-fold the annual N_2O production.

Keywords: autotrophic nitrogen oxidizers, biological technologies, greenhouse effect, nitrogen cycle, sponge media

1. Introduction

The ever-increasing nitrogen pollution in the environment is getting attention in recent years, particularly regarding the high warming-potential nitrous oxide (N_2O) and the discharge of the concerned ammonium (NH_4^+-N). First, the total emissions of N_2O reached 336.33 ppb (parts per billion) [1], accounting for 6.2% of GHGs (greenhouse gases), and its heat-capturing capacity is approximately 298 times higher than that of carbon dioxide (CO_2), calculated over a 100-year period. The escalating rate of GHG emissions will accelerate global warming, leading to a projected 1.5°C temperature increase before 2030. N_2O is also a primary contributor to ozone depletion, along with chlorofluorocarbons (CFCs), which amplifies the impact on the extreme climate [2]. Furthermore, nitrogen oxides (NOx, including NO and NO_2) readily dissolve in water vapor, leading to the formation of acid rain, which releases heavy metals from soil, indirectly poisoning various organisms and causing ocean acidification. Recent findings indicate that N_2O accumulation results from the quantitative release of marine and terrestrial environments, influenced by the three major human activities of agriculture, chemical factories and wastewater treatment plants (WWTP) [3].

Ammonium (NH_4^+-N) is another concerned compound, primarily originating from sources, such as animal husbandry, industrial and domestic sewage. Its impact is multifaceted, affecting gas, liquid and solid phases. The Environmental Protection Agency of Taiwan (Taiwan EPA) has set the regulatory limits of NH_4^+-N and total nitrogen (TN) in the discharge, with the limits being lower than 30 and 35 milligrams nitrogen per liter (mgN/L), respectively. These regulations are set to be enforced in 2024. However, many industries and sewage treatment plants face challenges in treating the wastewater containing nitrogenous compounds to meet the EPA regulation. Moreover, NH_4^+-N in the systematic environment can convert to pungent ammonia (NH_3) (pKa of $NH_4^+/NH_3 = 9.25$) under high pH conditions. NH_3 is a controlled component under the Convention on Long-Range Transboundary Air Pollution (CLRTAP), as outlined in the Gothenburg Protocol [3].

Various methods have been employed to eliminate nitrogen pollutants, such as NH_4^+-N and N_2O, including ion exchange resins, physical/biological adsorption and biological filtration [4], as well as thermal catalytic cracking and photocatalytic decomposition [5]. However, these treatment technologies often require significant initial investments and ongoing costs for consumable replacements. In recent years, biological treatment technologies with lower costs have been widely applied for the transformation of NH_4^+-N in diverse environmental settings, encompassing processes such as nitrification, denitrification and other nitrogen-removal procedures [6]. Among them, the reduction of N_2O in the last stage of the denitrification process has become the prevailing method for N_2O elimination in the aquatic system. Achieving a N_2O conversion rate of over 70% is possible when there is an ample carbon source available in the system. However, in low-carbon environments, the release of N_2O becomes unfavorable for denitrifiers relying on high-carbon-to-nitrogen (C/N)-ratio food sources. Hence, autotrophic nitrogen oxidizers, including ammonia-oxidizing microorganism (AOM) and complete nitrifying bacteria (complete ammonia oxidizer, comammox), are potentially valuable contributors to reducing the residual N_2O level in the atmosphere. These autotrophic nitrogen oxidizers offer the advantages of low cost and energy consumption.

Numerous chemical reactions based on the nitrogen cycle [3] have been identified for NH_4^+-N removal, encompassing a total of 14 reactions. However, most of these reactions have primarily been investigated in laboratory-scale systems. In Taiwan, the aerobic activated sludge tank has emerged as a popular NH_4^+-N removal system in the water treatment plants. Nevertheless, meeting the regulations set by the Environmental Protection Agency to reduce the total nitrogen requirement to 35 mgN/L by the year 2024 poses a significant challenge. Furthermore, the complete removal of NH_4^+-N and nitrate (NO_3^--N) in urban sewage with low NH_4^+-N concentration (<50 mgN/L) and wastewater with a low C/N ratio is problematic due to slow-growth rate of microorganisms and insufficient carbon sources. Consequently, treatment processes based on the mechanisms of autotrophic microbes have been proposed as valuable tools for the elimination of nitrogenous compounds in the wastewater.

In this chapter, we will explore the chemical substances and functional microorganisms that play pivotal roles in the nitrogen cycle. Additionally, we will investigate the utilization of two sponge-based biological systems to enhance the growth rate of the slow-growth functional microbes. Specifically, we will focus on the application of autotrophic nitrogen oxidizers for mitigating residual nitrogen pollutants in the environment.

2. Impact of nitrogen pollution on the environment

Water pollution in Taiwan is primarily attributed to the discharge of pollutants from factories, livestock excretion and domestic sewage, accounting for 34.08, 2.75 and 63.17% of the total annual discharge of 3.2 billion cubic meters, respectively. Nitrogenous compounds originate from petrochemical industry (45%), high-tech industry (23%), pig manure, urine wastewater (16%) and domestic sewage (14%). A continuous release of nitrogen pollutants into the atmosphere and water bodies can lead to eutrophication and hypoxia in aquatic ecosystems (resulting from liquidous NH_4^+-N), detrimentally affecting native plant species (associated with liquidous NO_3^--N), causing acid rain (resulting from gaseous NO_2), contributing to global warming (related to gaseous N_2O) and posing various other environmental challenges. The concentration of NH_4^+-N in the petrochemical industrial wastewater ranges from 10 to 300 mgN/L (with a C/N ratio of approximately 3) [6], while domestic sewage typically contains 18.8 ± 5.71 mgN/L of NH_4^+-N and 20.4 ± 6.38 mgN/L of total nitrogen (TN). The discharge of such sewage into natural water bodies accounts for 41.6% of national river pollution, with 7.2% classified as severe pollution (NH_4^+-N > 3.0 mgN/L) (National Environmental Water Quality Monitoring Annual Report in 2022). These findings have further implications for groundwater systems, where 42.0% of regions exceed regulatory limits. Particularly, the Taipei Basin (<0.01 ~ 8.89 mgN/L) and the Jianan Plain (<0.01 ~ 8.76 mgN/L) exhibit the most severe contamination levels (statistical data obtained from the National Environmental Water Quality Monitoring System in 2022).

On the other hand, N_2O, a well-known greenhouse gas, has garnered significant attention due to in contribution to global warming, reaching 336.33 ppb in December 2022 [1]. Approximately 4 teragrams (Tg) per year of nitrogen is released into the atmosphere from oceanic sources, while terrestrial source contributes around 12 Tg per year of nitrogen [7]. Human activities account for 40% of total greenhouse gas emission, with specific sectors making varying contributions [8]. Agricultural soil management is responsible for 74% of emissions, wastewater treatment for 6%, stationary combustion for 5%, chemical production and other product uses for 5%, manure management for 5%, transportation for 4% and other activities for 1% [9]. In terms of industrial processes, the largest amount of N_2O is produced from nitric acid (HNO_3), with an annual emission of about 400 metric tons [10]. Biological nitrogen-removal systems, including the activated sludge system (0.06%), nitrification (2.7–9%) [11], partial nitrification (nitritation), anaerobic ammonia oxidation (anammox), nitritation-anammox procedure (1.3–2.2%) [12], nitrifier denitrification, denitrification (0.6–1.9%) [13] and nitrification-denitrification process (1.9–8.5%) [14], have been identified as potential sources of N_2O emission [15]. Nearly 70% of these emissions are attributed to the NH_4^+-N oxidation process, resulting in a nitrogen conversion of 27% with equivalent to 600 parts per million by volume (ppmv) [16]. In the family of microbes involved in the nitrogen cycle, aerobic ammonia-oxidizing bacteria (aerAOB) have been found to release higher amounts of N_2O compared to aerobic ammonia-oxidizing archaea (aerAOA) and comammox bacteria [17].

3. Elimination of nitrogen pollutants from the surroundings

Nitrogen compounds, characterized by low molecular weight and high reactivity, have the ability to rapidly disperse into gas phase (atmosphere), liquid phase (various water bodies) and solid phase (soil or sediment), posing biological hazards and contributing to global warming. Many countries or organization, including the United States, Japan and the European Union, have implemented regulations to control the nitrogen concentrations in wastewater discharge, with limits set at less than 60 mgN/L. In Taiwan, the regulations will further restrict the total nitrogen content in public sewer systems to below 35 mgN/L by 2024. Of particular concern is N_2O, which possesses a high greenhouse potential (GHP) and is approximately 298-fold more potent than CO_2 in terms of its heat-trapping capacity. The significant impact of N_2O on global temperature rise cannot be ignored. The 2015 Paris Agreement, signed by 200 countries, aims to mitigate the rate of global warming and limit the temperature increase to within 2°C by the end of the twenty-first century. This collective effort reflects the global commitment to combat the effects of N_2O and other greenhouse gases on climate change.

The commonly physical and chemical methods are employed for the removal of NH_4^+-N from wastewater. These methods include air stripping, ion exchange, reverse osmosis, electrodialysis and breakpoint chlorination, among others. They enable the efficient conversion or recovery of different forms of ammonium. However, these techniques are often associated with high operational costs and the challenge of disposing of secondary compounds, which limits their economic viability. In the case of N_2O reduction, thermocatalytic methods [18] and photocatalytic methods [5] have been utilized for N_2O decomposition. The use of thermocatalysis dates back to the 1950s [19], and it involves the utilization of various media such as metals, reducing oxides and zeolites [18]. Photocatalytic methods commonly employ zerovalent zinc [20]. Despite their effectiveness, physicochemical techniques are rarely used for N_2O reduction in wastewater treatment plants. This is primarily attributed to the high levels of dissolved oxygen and the low concentration of N_2O typically found in the water field [3].

Considering the aforementioned challenges, the environment-friendly and cost-effective biological treatment technologies present a promising approach for addressing the residual amounts of NH_4^+-N or N_2O in the environment. In the case of wastewater treatment, the selection of the appropriate biological treatment system depends on the prevailing C/N ratio. A nitrification-denitrification system is suitable for high C/N ratios, whereas a nitritation-anammox process is more effective for low C/N ratios [21]. In terms of N_2O reduction, two main reaction pathways are commonly considered. The first pathway describes denitrification, where N_2O is reduced to N_2 as the final stage of the process [22]. The second pathway involves NO_3^--N ammonification [23]. Our research team has also been exploring an alternative elimination pathway involving oxidation [24]; however, the precise mechanism underlying NH_2OH generation in this process is still not fully understood. Further investigations are needed to unravel this key aspect.

4. Pathway of nitrogen transformation

Nitrogen is a vital element in the biosphere, playing a crucial role in atmospheric composition and the metabolic processes of living organisms. The nitrogen cycle encompasses various catabolic and anabolic reactions that drive the transformation of nitrogen compounds [25]. Five catabolic reactions include nitritification, nitratification, denitrification, dissimilatory nitrate reduction and anaerobic ammonium oxidation. These processes involve the conversion of nitrogenous compounds to different forms, facilitating their cycling within ecosystems. On the other hand, three anabolic reactions encompass ammonium uptake, assimilatory nitrate reduction and nitrogen fixation, which are responsible for the incorporation of nitrogen into organic molecules and the production of nitrogenous compounds essential for life processes. Additionally, ammonification is a crucial step in the nitrogen cycle, where organic nitrogen is converted back to ammonium. Understanding the intricate pathways of nitrogen transformation is essential for comprehending the dynamics of nitrogen cycling in various environmental systems.

The fundamental mechanisms of nitrogenous oxidation and reduction are nitrification and denitrification. Nitrification can be categorized into autotrophic nitrification and heterotrophic nitrification by different microbial communities. The former is carried out by aerobic autotrophic ammonia and nitrite oxidizers, and the latter is catalyzed by fungi as well as heterotrophic bacteria [26]. Autotrophic nitrification is a chemolithotrophic oxidation of ammonia to nitrate under strict aerobic conditions and conducted in two sequential oxidative stages: ammonia oxidation and nitrite oxidation. The yield of cells per unit of ammonia oxidizer (AOB) as the genus *Nitrosomonas* is approximately 0.15 mg cells/mg NH_4^+-$N_{oxidized}$, while for nitrite-oxidizing bacteria (NOB) such as the genus *Nitrobacter*, it is around 0.02 mg cells/mg NO_2^--$N_{oxidized}$. Oxygen consumption during these reactions is estimated to be 3.16 mg O_2/mg NH_4^+-$N_{oxidized}$ and 1.11 mg O_2/mg NO_2^--$N_{oxidized}$, respectively. Additionally, alkalinity in the form of 7.07 mg $CaCO_3$/mg NH_4^+-N is required for ammonium oxidation to maintain pH stability in the system.

Denitrification is the reduction of the oxidized nitrogen compounds (NO_2^--N and NO_3^--N) to dinitrogen (N_2) through the intermediate production of nitrogen oxide (NO) and N_2O. This transformation occurs via three distinct pathways, including respiratory denitrification, aerobic denitrification and lithoautotrophic denitrification. In the respiratory denitrification, heterotrophic microorganisms use nitrite and/or nitrate as electron acceptors, while organic matter serves as the carbon and energy source in the absence of oxygen [27]. In environmental biotechnology applications, a variety of electron donors and carbon sources, such as methanol, acetate, glucose, ethanol and others, can be used to facilitate respiratory denitrification. Next, aerobic denitrification is complete denitrification occurring at high dissolved oxygen (DO) concentration, and heterotrophic organisms are responsible for corespiration of nitrate and oxygen and they are widespread in the environment [28]. Third, autotrophic denitrifiers catalyzed the lithoautotrophic denitrification using inorganic sulfur compounds, hydrogen or ammonia as electron donors [28]. These specialized microorganisms play a crucial role in the removal of nitrogen compounds in specific ecological niches.

Nitrous oxide (N_2O) serves as a common intermediate in various nitrogen treatment systems. Within the nitrogen cycle, N_2O is primarily produced through three metabolic mechanisms, namely (1) oxidation of hydroxylamine (NH_2OH) [30], (2) nitrifier denitrification [31] and (3) anoxic nitrite reduction. First of all, NH_2OH oxidation plays a crucial role in ammonium oxidation and is a significant reaction leading to N_2O production. This process is catalyzed by hydroxylamine dehydrogenase (HAO) enzymes [32]. Two pathways have been identified for NH_2OH oxidation: (a) NH_2OH is first oxidized to NOH, which is subsequently chemically converted to N_2O [33]; (b) NH_2OH is first oxidized to NO, followed by enzymatic reduction to N_2O mediated by cytochrome c554 (cyt c554) [34].

The second is nitrifier denitrification, and the main player as *Nitrosomonas europaea* and other AOBs can reduce NO_2^- to NO, N_2O or N_2 in the absence of oxygen [35].

Two enzymes involved in this reaction are nitrite reductase (NIR) [36] and nitric oxide reductase (NOR) [37]. NIR enzymes catalyze the reduction of NO_2^--N to NO and subsequently, NOR enzymes facilitate the reduction of NO to N_2O [38]. Studies have shown that *N. europaea* lacking NIR enzyme produced four fold higher amounts of N_2O compared to the wild type with NIR enzyme, indicating the role of NIR in supporting HAO enzymes to enhance nitrification activity [39]. Notably, strains lacking NOR enzymes did not exhibit a significant effect on N_2O production [37]. These research findings suggest that NIR enzymes can support the function of HAO enzymes to raise the nitrification activity under the sufficient electron source. Therefore, nitrifier denitrification is not the main source of N_2O emission under normal situation of microbial growth. Overall, the metabolic pathways involving NH_2OH oxidation and nitrifier denitrification contribute to the production of N_2O within the nitrogen cycle. Further investigation into these mechanisms is necessary to better understand N_2O emissions and develop effective strategies for its mitigation.

On the other hand, the release of N_2O can occur under four different environmental conditions during the anoxic nitrite reduction. The first condition is N_2O accumulation due to the inhibition of nitrous oxide reductase (N_2OR) under DO concentration attaining 0.2–0.5 mg/L [40]. The second condition arises when N_2OR becomes inactive, interrupting the reduction of N_2O to N_2 at low pH levels [23]. The third condition occurs when there is an insufficient electron source that resulted from a low biodegradable organic load [41]. Lastly, nitrite (NO_2^--N) as the electron acceptor is more prone to induce the N_2O accumulation catalyzed by NIR/NOR enzymes, with a conversion rate of 55% per N transformation, compared to nitrate (NO_3^--N) at 0.8%/N transformation [42]. It is important to note that anammox bacteria and nitrite oxidizers are unlikely contributors to N_2O production, as the pathways for their potential generation of N_2O have been elucidated. Instead, four factors, including microaerobic environment, insufficient electron source, NO_2^--N accumulation and acidification, likely stimulate ammonia oxidizers and denitrifiers to produce N_2O in wastewater treatment systems. Furthermore, NO_2^--N accumulation and acidification also promote abiotic decomposition processes that contribute to N_2O emission.

To further eliminate the presence of high-GHP-N_2O, two biological reduction mechanisms have been identified: (1) N_2O reduction during denitrification and (2) NO_3^--N ammoniation. During denitrification, nitrous oxide reductase (N_2OR) catalyzes the reduction of N_2O to N_2 [43]. In addition, studies have revealed the growth and activity of *Rhodobacter capsulatus* and *Wolinella succinogens* in the presence of high N_2O concentration [44, 45]. However, N_2O becomes the final product of denitrification at low C/N ratio in an influent, making it challenging to further initiate the reduction of N_2O to N_2. In terms of NO_3^--N ammoniation, *Bacillus vireti* utilize N_2O oxidized to NO_x by activating the NOS operon under anaerobic condition, while simultaneously synthesizing microbial cells.

5. Biological technologies based on the nitrogen cycle

The biological nitrogen cycle encompasses 14 currently known biochemical conversion mechanisms, which can be broadly categorized into nitrification, comammox, denitrification, anammox, nitrate assimilation, respiratory ammonification (dissimilatory nitrate reduction to ammonia (DNRA)) and nitrogen fixation. Nitrification is a well-established process employed in wastewater treatment plants, where NH_4^+-N is sequentially oxidized to NO_3^--N via three steps with the intermediates of NH_2OH and NO_2^--N. In contrast, comammox performs the direct one-step oxidation of NH_4^+-N to NO_3^--N [46]. Furthermore, NO_3^--N serves as the inducer for initiating denitrification, which involves a four-stage reduction of NO_3^--N to N_2 with the intermediates of NO_2^--N, NO and N_2O.

In contrast to high-carbon-demanding denitrification, autotrophically anaerobic ammonia oxidation (anammox) has attracted significant attention due to its low energy consumption and minimal sludge production. Anammox is an energetically favorable reaction that utilizes NH_4^+-N and NO_2^--N/nitrogen oxide (NO) as electron donors and acceptors to yield gaseous nitrogen. Furthermore, anammox organisms utilize CO_2 as the sole carbon source for cellular material synthesis [47]. Notably, hydrazine (N_2H_4) plays a crucial role as an electron donor in the conversion of NO_2^--N to NH_2OH in the anammox process, distinguishing it from other nitrogen removal processes.

To achieve the comprehensive removal of nitrogen compounds, various biochemical reactions and their combinations have been applied. For instance, nitrification-denitrification process has been recommended for the wastewater containing a high C/N ratio, whereas nitritation-anammox system is suitable for the influent with a low C/N ratio. In the two-stage nitrification-denitrification process, organic matter in wastewater is initially degraded to lighten the inhibition of autotrophic nitrifiers. The resulting NH_4^+-N is then further oxidized to NO_3^--N by nitrification. NO_3^--N can be circularly used as electron acceptor for denitrification, leading to 70–90% nitrogen removal after the long-term operation. In comparison to two-stage nitrification-denitrification, a single reactor that combines the advantages of both reactions has been developed, known as the SHARON process (the acronym for Single reactor High activity Ammonium Removal Over Nitrite) [48].

In the case of nitritation-anammox, NH_4^+-N undergoes partial oxidation to NO_2^--N by supplying 75% of the required oxygen, as opposed to the complete oxidization of NO_{3-}-N. Subsequently, NH_4^+-N and NO_2^--N are reduced to N_2. The partial oxidation of NH_4^+-N is also known as partial nitrification [49], as it directly provides the necessary substrates for the anammox family without extra energy consumption. Throughout this process, two groups of autotrophic microbes work together to convert NH_4^+-N into N_2, making it well suited for the wastewater with low organic content. In comparison to the two-stage system, a single reactor is employed to facilitate the growth of both autotrophic aerobically and anaerobically ammonia oxidizers, which are responsible for the transformation of NH_4^+-N to N_2. This process is commonly referred to as CANON (the acronym for Completely Autotrophic Nitrogen removal Over Nitrite) [50].

To simplify the understanding of the system's functionality, we focus on the characteristics of an ammonia oxidizer such as the genus *Nitrosomonas*. One system that controls the activity of nitrification and denitrification through the regulation of oxygen is known as the OLAND process (Oxygen-Limited Autotrophic Nitrification and Denitrification) [51]. Another system, referred to as the NO_x process [52], operates by regulating the levels of NO_x (NO/NO_2) to facilitate nitrification and denitrification. Additionally, the archaeal family can anaerobically oxidize methane (CH_4) coupled with NO_3^--N reduction, known as N-damo [51]. In comparison to anammox, N-damo achieved a further reduction of 0.19 mM CH_4 while utilizing 1 mM NH_4^+-N [53]. On the other hand, aerobic deammonification directly converts NH_4^+-N to N_2 and NO_2^--N via NH_2OH, although the detailed mechanism of this process is not yet well understood [54].

Recently, significant attention has been given to the production of GHP-N_2O through four reactions involved in the nitrogen cycle, including NH_2OH oxidization [30], nitrifier denitrification [31], comammox [46] and NO_2^--N reduction [37]. During NH_2OH oxidation, it is believed that hydroxylamine oxidase (NH_2OH oxidase, HAO) or nitric oxide reductase (NO reductase, NOR) present in microbial cells catalyzes the oxidation or reduction pathways for N_2O formation [30]. Genus *Nitrosomonas*, in the process of nitrifier denitrification, utilizes NH_4^+-N or nitrogen oxides to produce N_2O under anoxic condition [55]. The comammox reaction, facilitated by the genus *Nitrospira*, also leads to the release of N_2O [56]. The final pathway is NO_2^--N reduction, which occurs under conditions of high dissolved oxygen [56], low pH [23], insufficient organic loading [43] and NO_2^--N in replacement of NO_3^--N as electron acceptor [42]. The main mechanism of N_2O reduction is the reduction of N_2O to N_2 catalyzed by N_2OR enzyme (clade II nosZ) in denitrifiers [57]. In addition, there are still unclear mechanisms of N_2O elimination, including the co-metabolism of NO_3^--N ammonification [23] and N_2O nitrification [24].

6. Key microbes involved in the nitrogen cycle

The nitrogen cycle involves the participation of six prominent groups of microorganisms responsible for 14 biological reactions. These groups are aerobic ammonia-oxidizing bacteria (aerAOB), aerobic ammonia-oxidizing archaea (aerAOA), anaerobic ammonia-oxidizing bacteria (anAOB or anammox bacteria (AMX)), nitrite-oxidizing bacteria (NOB), denitrifying microbes (DENer) and nitrogen-fixing bacteria (NFB).

The first group is aerobic ammonia-oxidizing bacteria (aerAOB), and it uses ammonia monooxygenase (AMO) and hydroxylamine dehydrogenase (HAO) to catalyze the oxidation of NH_4^+-N to NO_2^--N via NH_2OH. This family comprises six genera of *Nitrosomonas*, *Nitrosolobus*, *Nitrosovibrio*, *Nitrosospira*, *Nitrosococcus* and Candidatus *Nitrosoglobus* [58] within two bacterial phyla of β- and *γ-proteobacteria* [25]. Among them, the genus *Nitrosomonas* is not only an obligate autotrophic nitrifier but can also act as a denitrifier, reducing NO_2^--N using hydrogen (H_2) as an electron donor [31]. Furthermore, the oxidation of NH_2OH to NO is initially catalyzed by HAO enzyme, and then NO is converted to NO_2^--N by nitric oxide oxidoreductase (NOO) [56]. However, the oxidation of NH_2OH to N_2O occurs with NO_2^--N as the electron acceptor in the absence of oxygen [59].

The second group is aerobic ammonia-oxidizing archaea (aerAOA), and it catalyzes the similar mechanism of NH_4^+-N oxidation as aerobic ammonia-oxidizing bacteria (aerAOB). However, there is a distinction in the process: the intermediate of NO in the NH_2OH oxidation is rapidly consumed, and no free NO is released to the atmosphere. This family encompasses 10 genera of *Nitrosoarchaeum*, *Nitrosopumilus*, *Cenarchaeum*, *Nitrososphaera*, Candidatus *Nitrosocaldus*, Candidatus *Nitrosotalea* [60], Candidatus *Nitrosotenuis* [61], Candidatus *Nitrosopelagicus*, Candidatus *Nitrosocosmicus* [62] and Candidatus *Nitrosomarinus* [63] within the phylum *Thaumarchaeota*. These archaea are prevalent in the ocean and interact with other team players in the system, contributing to one third of total N_2O emission.

The third group is anaerobic ammonia-oxidizing bacteria (anAOB or AMX), and it performs the reduction of NH_4^+-N and NO/NO_2^--N to N_2. Three families of anammox microbes with distinct biogeographical distributions have been identified: freshwater Candidatus *Brocadiaceae* (including the genera *Brocadia*, *Kuenenia*, *Anammoxoglobus* and

Jettenia) [64], marine Candidatus *Scalinduaceae* (represented by the genus *Scalindua*) [65] and marine Candidatus *Bathyanammoxibiaceae* [66] in the order Candidatus *Brocadiales* within the phylum *Planctomycetes* [67]. Notably, *Anammoxoglobus propionicus* exhibits the ability to reduce NO_2^--N by simultaneously utilizing NH_4^+-N and propionate as electron donors [68]. Furthermore, *Kuenenia stuttgartiensis* is capable of performing dissimilatory NO_3^--N reduction to NH_4^+-N (DNRA) [69]. Additionally, this group demonstrates the capacity for carbon fixation under anaerobic condition [26], making it advantageous for GHP-CO_2 elimination applications.

The fourth group is nitrite-oxidizing bacteria (NOB) and it conducts the oxidation of NO_2^--N to NO_3^--N. This family includes seven genera: *Nitrobacter* in the phylum *α-proteobacteria*, Candidatus *Nitrotoga* in the phylum *β-proteobacteria*, *Nitrococcus* in the phylum *γ-proteobacteria*, *Nitrospira* in the phylum *Nitrospirota*, both of *Nitrospina* and Candidatus *Nitromaritima* in the phylum *Nitrosponota* and *Nitrolancea* in the phylum *Thermomicrobiota* [70, 71]. Notably, the widely distributed *Nitrospira* are further divided into canonical nitrite-oxidizing *Nitrospira* (canonical-*Nitrospira*) and comammox-*Nitrospira* [72]. In comparison of canonical-*Nitrospira*, comammox-*Nitrospira* catalyzes the complete oxidization of NH_4^+-N to NO_3^--N. For example, Candidatus *Nitrospira inopinata* exhibits higher affinility of NH_4^+-N than ammonia oxidizer under the limited NH_4^+-N condition [73]. Moreover, Candidatus *Nitrologa* has been discovered in marine environments and demonstrates tolerance to high salinity [72].

The fifth group is denitrifying microbes (DENer), and they possess the ability to reduce nitrogen oxides (such as NO_3^--N, NO_2^--N, NO and N_2O) to N_2 under anaerobic, micro-aerophilic and occasionally aerobic conditions. This diverse family can be categorized into heterotrophs, autotrophs and mixotrophs based on their energy source. While heterotrophic denitrifier commonly utilizes organics as electron donor, autotrophic denitrifier (AuDen) primarily uses H_2, reduced inorganic sulfur compounds (RISCs, such as S^0, S^{2-} and $S_2O_3^{2-}$), sulfite (SO_3^{2-}), thiocyanate (SCN^-), iron oxides (e.g., iron disulfide (FeS_2), Fe^{2+} and Fe^0 or zerovalent iron (ZVI)) and trivalent arsenic (As^{3+}). The autotrophic families belong to various phyla including α -, β-, γ- and *ε-proteobacteria* [74]. It is noteworthy that certain nondenitrifying microbes possess the N_2OR enzyme that can directly utilize the residual N_2O in the environment as a source of energy and nutrients, including the genera *Anaeromyxobacter*, *Dyadobacter*, *Gemmatimonas*, *Ignavibacterium*, *Melioribacter* and *Pedobacter*. They potentially play a role in the elimination of GHP-N_2O [75]. Lastly, nitrogen-fixing bacteria (NFB) catalyze the reduction of N_2 to NH_3 through the process of nitrogen fixation. They are widely distributed in various environemts, including the phyla *Proteobacteria*, *Chlorobi*, *Firmicutes* and *Cyanobacteria*, and three methanogenic archaea of the genera *Methanosarcina*, *Methanococcus* and *Methanothermobacter* [76].

7. The fixation of carbon dioxide by autotrophic nitrogen families

Carbon dioxide (CO_2) is an essential element for living organisms; however, its concentration was calculated to be 421.00 ppm (parts per million) in March 2023 [1], contributing to the global warming. Microbes are the key contributors for biological CO_2 elimination by utilizing six different pathways for cell or carbohydrate syntheses. Six known pathways of CO_2 fixation are Wood-Ljungdahl pathway (W-L), 3-Hydroxypropionate 4-hydroxybutyrate cycle (3HP-4HB),

Calvin-Benson-Bassham cycle (CBB), 3-Hydroxypropionate cycle (3-HP), Reductive tricarboxylic acid cycle (rTCA) and Dicarboxylate 4-hydroxybutyrate cycle (DC-4HB) [77]. In the context of autotrophic CO_2 assimilation, five distinct groups of nitrogen-related microbes are involved in the four above-mentioned pathways. First, aerobic ammonia-oxidizing archaea including mesophilic *Crenarchaeota* and thermophilic *Thaumarchaeota* prefer to use their respective modified versions of 3-Hydroxypropionate 4-hydroxybutyrate cycle (3HP-4HB) known as the Crenarchaeal HP/HB cycle and Thaumarchaeal HP/HB cycle [78]. These pathways enable them to assimilate CO_2 and carry out NH_4^+-N oxidation simultaneously. Second, anaerobic ammonia oxidizers (anammox bacteria) employ the Wood-Ljungdahl pathway for CO_2 assimilation during their unique anaerobic ammonia oxidation process [64]. Third, the Calvin Benson Bassham cycle (CBB) is present in ammonia-oxidizing bacteria [79] as well as in the four genera of nitrite oxidizers, namely *Nitrobacter*, *Nitrococcus*, *Nitrotoga* and *Nitrolancea* [80]. These organisms utilize the CBB cycle to fix CO_2 and perform their respective oxidation processes. Fourth, nitrite-oxidizing *Nitrospira* and autotrophic-denitrifying *Thiobacillus denitrificans* have been found to involve in Reductive tricarboxylic acid (rTCA) cycle [74]. This pathway allows these organisms to fix CO_2 while carrying out NO_2^--N oxidation or denitrification. It is worth noting that nitrite-oxidizing *Nitrococcus* and *Nitrospina* have higher potential for CO_2 utilization in marine environments compared to ammonia-oxidizing archaea and bacteria, particularly during the early exponential phase of microbial growth. Conversely, ammonia-oxidizing *Nitrosomonas* demonstrate rapid rate of cell synthesis in both late exponential and stationary phases [81]. In summary, autotrophic nitrogen-functional microbes possess the remarkable ability to utilize nitrogen compounds and CO_2 as energy source and cell synthesis. This capability not only contributes to the reduction in global warming but also aids in the removal of nitrogenous pollutants from the environment.

8. Cultivation systems of nitrogen-functional microbes

The cultivation of nitrogen-functional microbes relies on providing suitable energy sources for their growth. This includes nitrogen sources, alkalinity (typically $NaHCO_3$), buffer (KH_2PO_4 and Na_2HPO_4), nutrients and trace elements [82]. Different nitrogen sources are utilized depending on the specific group of microbes being cultivated. For instance, NH_4^+-N, NO_2^--N, NH_4^+-N/NO_2^--N, NO_2^--N/NO_3^--N and N_2O were, respectively, used for ammonia oxidizers (AOA and AOB), nitrite oxidizer, anammox, denitrifier and N_2O-utilizing microbes. In addition to nitrogen sources, essential nutrients, such as calcium, magnesium and iron, are provided through $CaCl_2$, $MgSO_4$ and $FeSO_4$. In the case of ammonia oxidizers carrying out partial nitrification, Na_2SO_4 is added as a supplement. Trace elements, which are crucial for microbial growth, consist of $CuSO_4$, $ZnSO_4$, $MnCl_2$, $NiCl_2$, $CoCl_2$, $NaMoO_4$, $NaSeO_4$, $NaWO_4$, Na_2-EDTA and H_3BO_4. To avoid Na_2-EDTA and H_3BO_4 from serving as carbon source for the growth of heterotrophic bacteria, they are excluded from the trace elements for cultivation of N_2O-utilizing microbes. These cultivation systems aim to provide the necessary nutrients and conditions for the successful growth of nitrogen-functional microbes, enabling their study and potential application in various nitrogen cycling processes.

The cultivation system's design plays a crucial role in the successful enlargement of nitrogen-functional microbes. An important factor to consider is the choice of a suitable habitat for their growth. In this regard, the downflow hanging sponge (DHS) system utilizes a polyurethane sponge as a supporting material, as depicted in **Figure 1**. This sponge provides a three-dimensional (3D) space that facilitates the growth of microorganisms. When wastewater, containing microbes and foods, flows into the sponge, microbial cells are retained both inside and outside the sponge media. The unique microenvironment allows for the coexistence of aerobic and anaerobic nitrogen-functional microbes. Specifically, the surface of the sponge supports the growth of aerobic autotrophs responsible for nitritation, while deeper within the media, anaerobic microbes catalyze the anammox process [82]. Since its initial development in 1995, the DHS system has undergone several modifications, resulting in six different configurations of sponge setups [29]. The superiorities of the DHS are high biomass retention, long sludge retention, minimal sludge production and less energy consumption, particularly benefiting to cultivate the slow-growing autotrophic nitrogen-functional microbes. The combination of the polyurethane sponge as a support material and the unique microenvironment provided by the DHS system contributes to the successful cultivation of nitrogen-functional microbes, enabling their study and application in various nitrogen cycling processes.

The G1-type DHS reactor is a nonsubmerged fixed-bed reactor, illustrated in **Figure 2**. It consists of a closed rectangular column with a total volume of 2.5 L, while the working volume is 0.596 L, considering the 98.4% void ratio of the sponge material. Inside the reactor, 19 strips of triangular sponge (sized 2.8 × 2.8 × 4 cm) are arranged on two opposite inner walls, with a gap of 0.5 cm between each consecutive sponge. The height of the reactor column was 1 m, but the effective height was 2 m, as the sponge strips adhered on opposite walls were connected in series during the operation of the reactor. This design allows for enhanced contact between the wastewater and the sponge media, promoting efficient microbial growth and nutrient removal. Another cultivated system is the upflow submerged sponge (USS) reactor, shown in **Figure 3**. This reactor configuration provides an effective volume of 1.5 L within a 3-L column. The USS reactor employs a 1-L sponge volume as the attached media, creating a coexisted environment for suspended and biofilm-type microorganisms. The temperature control in the USS reactor is achieved through a cycled

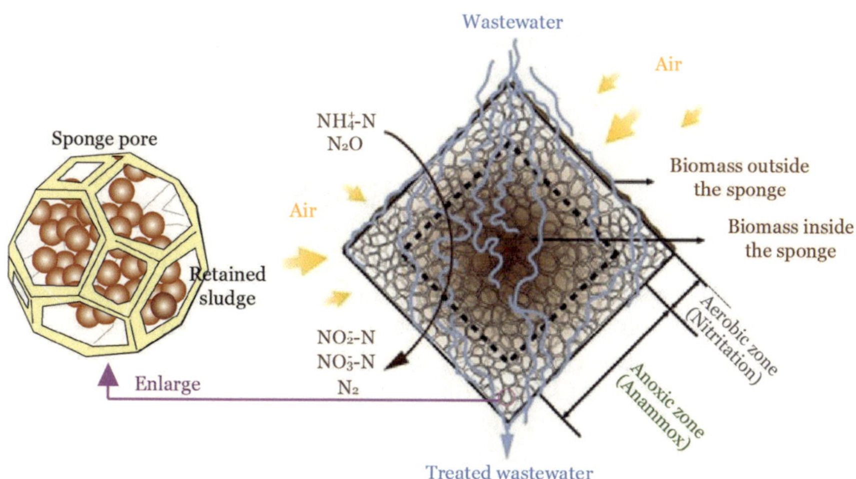

Figure 1.
Conceptual cross-sectional view of the cube-type downflow hanging sponge (DHS sponge) (modified from [29]).

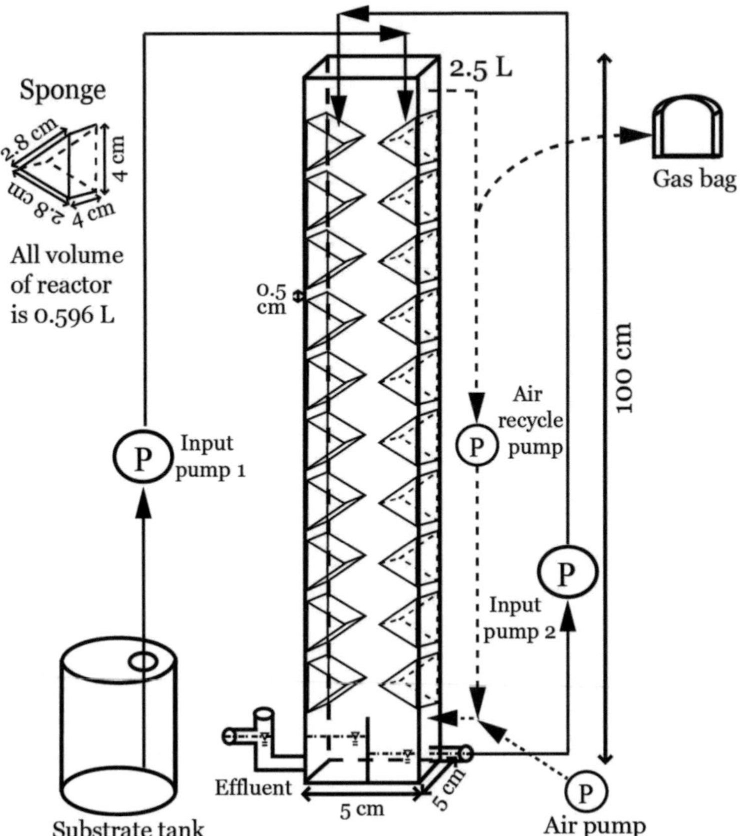

Figure 2.
Schematic diagram of G1-type downflow hanging sponge (DHS) system.

water system, ensuring optimal conditions for microbial activity and growth. These cultivation systems, namely the G1-type DHS reactor and the USS reactor, provide suitable environments for the growth of nitrogen-functional microbes. The well-designed arrangement of sponge media in these reactors allows for efficient nutrient utilization and microbial interactions, enabling the study and application of nitrogen cycling processes in wastewater treatment and environmental biotechnology.

9. Test procedure and analytical methods used in NH_4^+-N and N_2O oxidation

The G1-type DHS system was used for three processes that enhance NH_4^+-N transformation. First, a stepwise increase of NH_4^+-N concentration from 30 to 400 mg N/L was performed at eight different phases in the nitritation system, which corresponded to the nitrogen load of 0.47 to 6.42 kg N/m^3-day. Second, the potential of the anammox system was tested using four different parameters of cultivated temperature, inflow rate, substrate concentrations (including NH_4^+-N and NO_2^--N) and effluent recirculation. Third, the operational conditions of the complete nitrogen transformation in a single-type DHS reactor (namely CnDHS) were similar to those of the nitritation system. All systems were placed in the 30–35°C incubator. In addition, the

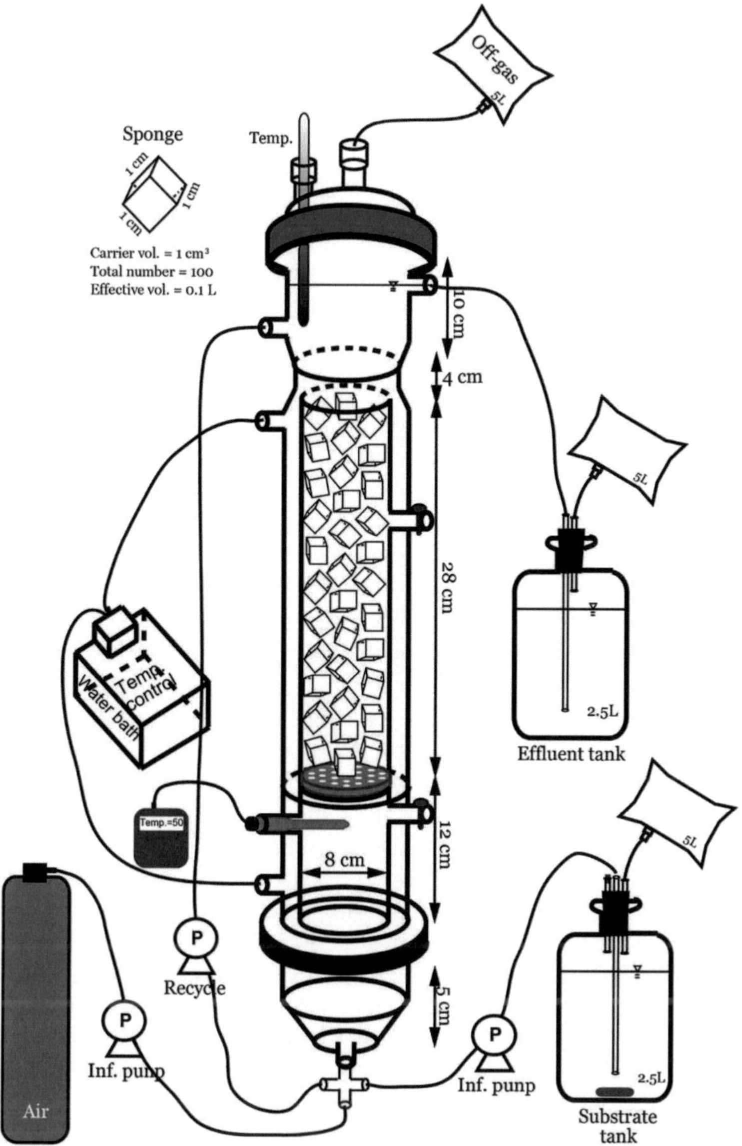

Figure 3.
Schematic diagram of upflow submerged sponge (USS) system.

low oxygen concentration inside both nitritation and CnDHS systems was controlled by adjusting the airflow rate from 1.5 to 16 L/day based on the partial pressure of oxygen in the reactor.

Two USS systems were used to improve the efficiency of anammox and N_2O oxidation. In the anammox reactor, the increase of nitrogen load was observed to take place from 9.60 to 38.4 mgN/L-day, along with the concentration increase of chloride from 160 to 1200 mg/L under a fixed HRT of 4.2 days. For N_2O oxidation, the rise of N_2O load was performed by increasing the substrate flow rate from 0.04 to 0.32 L/day (HRT shortened from 4 days to 0.5 days) under the fixed N_2O substrate of 25 mM in the liquid based on the 100% gaseous N_2O dissolved in the medium. The flow of air inside the reactor was controlled to adjust the desired oxygen concentration, and the exhaust gas from the reactor was collected using a gas bag. The microbial activity was further tested in the batch assays with different N_2O concentrations under the satisfactory oxygen conditions.

To monitor the performance of all systems, NH_4^+-N, NO_2^--N and NO_3^--N in influent and effluent were regularly measured using a colorimetric method (HACH, USA) and ion chromatograph (SH-120A, SHINE, New Zealand). The composition of off-gas was determined using gas chromatography (Shimadzu GC-8APT for O_2, GC-8AIT for N_2O, CO_2 and N_2). Theoretical DO concentration in the bulk liquid flowing on the sponge surface was computed from oxygen content in the gas phase according to modified Henry's equation, and the actual DO concentration in effluent was measured directly by a DO meter (YSI/Nanotech Inc., Japan). The concentration of the retained biomass in the sponge material was measured at the end of the operation, and the biomass was stored in a −20°C freezer for microbial clarification. DNA was first isolated using MOBIO PowerSoil DNA extraction kit, and microbial community and functional genes catalyzed nitrogen transformation were further analyzed by TOPO cloning kit and SybrGreen quantitative PCR (QuantStudio 1, ThermoFisher Scientific, USA) with the specific primer pairs.

10. Use of autotrophic N-removal processes for NH_4^+-N reduction

This section aims to assess the efficiency of NH_4^+-N conversion by comparing three autotrophic N-removal processes, including nitritation (also called partial nitrification), anammox and complete nitrogen transformation in a single-type reactor. The G1-type DHS is used for microbial enlargement and functional evaluation. The findings of the study are presented and discussed below.

In the operation of nitritation, the G1-type DHS reactor was subjected to a total of seven phases, with NH_4^+-N inflow rates ranging from 0.47 to 1.60 kgN/m^3-day. The hydraulic retention time (HRT) was fixed at 1.5 h, and the temperature was maintained at 30°C. The airflow rate, ranging from 2 to 16 L/day, was adjusted to control the oxygen content in the system. Microbes cultivated in the DHS exhibited high capability for NH_4^+-N oxidation, achieving rates of up to 1.92 kgN/m^3-day even at low oxygen condition. This performance surpasses those of a fixed-film bioreactor (0.58 kgN/m^3-day) [83] and submerged membrane bioreactor (1.30 kgN/m^3-day) [84] operating under sufficient oxygen supply. Furthermore, partial NH_4^+-N oxidation of 58.6% was attained at NH_4^+-N load of 3.24 kgN/m^3-day. This resulted in the production of 37.5% NO_2^--N and 4.0% NO_3^--N with an oxygen concentration of 0.40% O_2 (0.16 mg/L of DO) (see **Figure 4**) [85]. Similarly, a biofilm system demonstrated that partial nitrification with 50% NH_4^+-N conversion was attained at oxygen concentrations below 0.2 mg/L [86]. The growth rate of ammonium oxidizers was 2.56-fold faster than that of nitrite oxidizers under the DO concentration below 1.0 mg/L [87]. Similarly, the growth yield of *Nitrosomonas* sp. under oxygen stress as low as 1% was found to be 5 times higher compared to conditions with saturated DO conditions. The G1-type DHS reactor provided a high biomass concentration of 3.84 g volatile solids (VS)/L, enabling a nitrifying activity of 0.20 kg NH_4^+-N/kg VS-day. The activity of ammonia oxidizer in the DHS was comparable to those of the suspended growth-type reactors (0.17–0.29 kg NH_4^+-N/kg VS-day) [88], and higher than that of the biofilm-type system (0.08–0.10 kg NH_4^+-N/kg VS-day) [89]. However, GHP-N_2O production was detected at a level of 0.5% in the DHS, equivalent to 13% of the

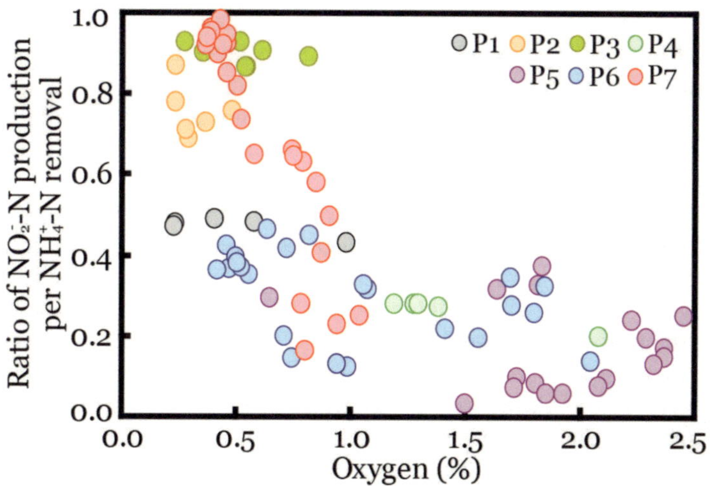

Figure 4.
Effect of O_2 in the downflow hanging sponge (DHS) for nitritation on the ratio of NO_2^--$N_{produced}$/NH_4^+-$N_{removed}$. Wherein, P1 ~ P7 is the data taken from phase 1 to phase 7 of the operation in the DHS.

oxidized NH_4^+-N under a gas-phase oxygen content of 0.4% (0.16 mg/L of DO) (see **Figure 5**). Similarly, N_2O production was observed from 10% of the oxidized NH_4^+-N under the O_2 concentration of 0.18 mg/L [30]. The analysis of microbial community revealed that 32.4% phylotypes closely related to *Nitrosomonas* sp. strain ENI-11 dominated in the DHS, while denitrifying genera of *Azoarcus* and *Bradyrhizobium* and nitrite-oxidizing *Nitrobacter* coexisted and participated in the nitrogen cycle [90]. The *amoA* gene encoded in the enzyme, which catalyzes ammonium oxidation, was used for determining the functional microbes, resulting in the phylotypes within the *Nitrosomonas europaea/Nitrosococcus mobilis* lineage being the key players in the nitritation in the DHS at low oxygen atmosphere. The images of fluorescence *in situ*

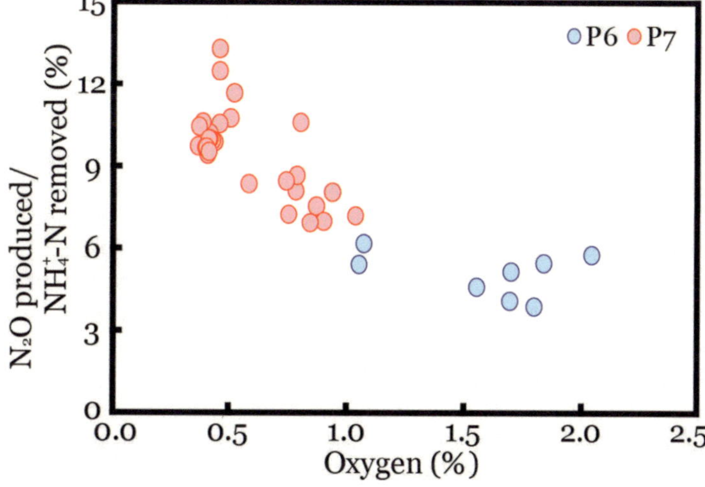

Figure 5.
Effect of O_2 in the downflow hanging sponge (DHS) for nitritation on the ratio of $N_2O_{produced}$/NH_4^+-$N_{removed}$. Wherein, P6 ~ P7 is the data taken from phase 6 to phase 7 of the operation in the DHS.

hybridization (FISH) demonstrated that 41% of β-proteobacterial ammonia oxidizers coexisted with 5.4% of *Nitrobacter* spp. within the bacterial community, accounting for 83% of the total population. Based on these findings, it can be concluded that the ammonia oxidizer as *Nitrosomonas* family was numerically dominant over nitrite oxidizer in the DHS reactor, facilitating the occurrence of nitritation at low oxygen supply. However, the presence of nitrite reductase, involved in N_2O production through nitrifier denitrification, was induced at low oxygen partial pressures [39].

The optimal proportion of NH_4^+-N and NO_2^--N for anammox reaction was achieved through first-stage nitritation. The DHS employed for the anammox process operated at a total nitrogen load ranging from 0.48 to 5.96 kgN/m^3-day with NH_4^+-N and NO_2^--N maintained at an equal proportion. The HRT was set between 0.7 and 2 h, and the reactor was operated at a temperature range of 30 to 35°C. The highest nitrogen-removal rate achieved in the DHS was 2.27 kg N/m^3-day, which surpassed the performance of other biofilm systems with the removal rates of 0.2 to 2.0 kgN/m^3-day [91, 92]. However, the nitrogen-removal rate in the DHS was lower than that reported for an upflow fixed-bed column reactor designed for highly enriched anammox [93]. The DHS exhibited a biomass concentration of 5.59 g VS/L within the sponge media, enabling anammox activity of 0.39 kgN/kg VS-day. Remarkably high removal efficiency of 95.4% was achieved at a loading rate of 1.94 kgN/m^3-day and HRT of 1.0 h, giving NO_2^--$N_{utilized}$/NH_4^+-$N_{removed}$ of 1.25 ± 0.080 and NO_3^--$N_{production}$/NH_4^+-$N_{removed}$ of 0.25 ± 0.042 [94]. Notably, no N_2O was detected in the DHS, highlighting the physiological capacity of anammox bacteria to suppress N_2O production [95]. Moreover, based on theoretical calculations, approximately 76% of the removed NH_4^+-N was converted to N_2 through the anammox reaction, while the remaining 24% was suggested to be consumed via NO_3^--N reduction processes, including assimilation and dissimilation, as well as denitrification [96]. The anoxic microenvironment within the sponge media of the DHS, as depicted in **Figure 1**, likely provided a reducing environment and limited carbon sources. Additionally, cell lysis resulting from microbial mortality during resting periods further contributed to the availability of organic matter. As discussed earlier, the co-occurrence of nitrate reduction or denitrification alongside the anammox reaction in the DHS led to higher nitrogen removal (95%) than other systems [93, 96–98]. The key microbial players in this community included anammox genera *Kuenenia* and *Anammoxoglobus*, ammonium-oxidizing genus *Nitrosomonas*, as well as denitrifying capability of the genera *Comamonas* [99] and *Diaphorobacter* [100]. Together, these microbial groups worked synergistically to reduce nitrogenous compounds and facilitate efficient nitrogen removal in the system.

On the contrary, when it comes to the USS system designed for treating high salinity wastewater with a low C/N ratio, a remarkable removal efficiency of 93.3% was attained at a nitrogen load of 38.4 mgN/L-day, even under a chloride (Cl^-) concentration of 300 mg/L. However, it should be noted that the increase of Cl^- concentration to 1200 mg/L resulted in an extended adaptation period of 1 month for the utilization of NH_4^+-N and NO_2^--N. Comparing the USS system to the DHS system used for treating fresh water, the USS system enlarged the main groups of anammox *Brocadia*, ammonia-oxidizing *Nitrosomonas*, canonical nitrite-oxidizing or comammox-functional *Nitrospira*. Additionally, denitrifying genera, such as *Denitratisoma*, *Acinetobacter*, *Pseudomonas* and *Comamonas*, were also observed in the bacterial community of the USS system. The activities of microbes involved in ammonia oxidation, comammox, anammox and denitrification were assessed using the functional indicators of *amoA*, *crenamoA*, *Nts-amoA*, *hszA*, *nirS* and *nirK* genes. As shown in **Figure 6**, the abundance of the former four genes notably increased after

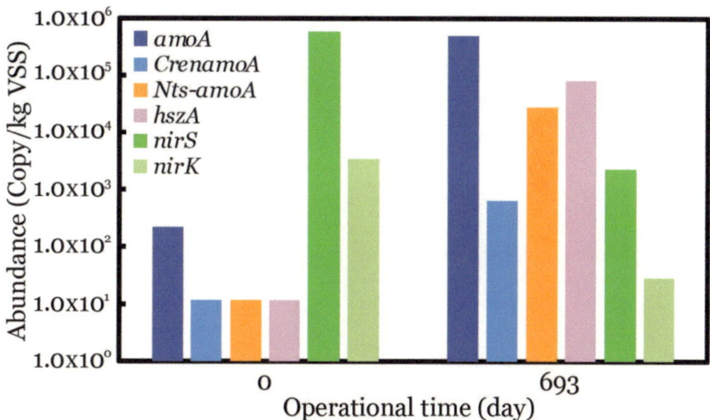

Figure 6.
Functional genes of microbes are involved in the conversion of NH_4^+-N and NO_2^--N in the. Upflow submerged sponge (USS) system for treating high-salinity wastewater. Wherein, amoA *and* crenamoA *stand for ammonium oxidation,* Nts-amoA *for Comamonas,* hszA *for anaerobic ammonia oxidation (anammox) reaction, and* nirS *and nirK for nitrite oxidation. In addition, the value below the detected limits of $2.32x10^1$ copy/kg VSS was used as $1.61x10^1$ copy/kg VSS.*

693 days of the cultivation in the USS system. In contrast, the presence of denitrifying *nirS* and *nirK* genes decreased over time. These findings suggest that anammox bacteria replaced the denitrifying microbes in facilitating the reduction of nitrogen oxides such as NO_2^--N and NO_3^--N. Additionally, the USS system demonstrated the dominance of slow-growing autotrophic nitrifiers harboring the *amoA* gene.

To optimize the synergy between nitritation and anammox, a single reactor capable of completely converting NH_4^+-N to N_2 was implemented in the DHS. Initially, the slow-growing and environmentally sensitive anammox bacteria were cultivated within the DHS, followed by the colonization of enlarged aerobic ammonia oxidizers coated on the outer surface of the sponge media. The DHS operation involved varying NH_4^+-N loads from 0.30 to 2.42 kg N/m^3-day, while maintaining limited oxygen levels controlled by airflow of 1.5 to 5.4 L/day. Remarkably, the maximum nitrogen-removal rate reached 1.53 kgN/m^3-day, surpassing the performance of suspended sludge system (0.2 kgN/m^3-day) [101] and biofilm-type reactor (1.5 kg N/m^3-day) [89]. Furthermore, this system demonstrated stable autotrophic nitrogen removal, even at an HRT as short as 2 h, in contrast to other processes requiring much longer HRTs (up to 10 h). Notably, an impressive efficiency of 84.8% was attained at NH_4^+-N load of 1.51 kgN/m^3-day, giving 84.5% of N_2 production alongside 8.0% NH_4^+-N and 7.5% of nitrogen oxides. The precise adjustment of oxygen content in the DHS proved crucial in controlling the proportion of NO_3^--N and N_2 production. In **Figure 7**, it is evident that an oxygen concentration of 1.0% serves as the critical threshold for distinguishing the dominant reaction pathway between anammox and nitrification, as indicated by the ratio of NO_3^--N/(NO_2^--N + NO_3^--N). Anammox bacteria predominantly catalyze the production of NO_3^--N when the O_2 content falls below 1%, whereas complete nitrification, driven by the faster growth rate of nitrite oxidizers compared to ammonia oxidizers, occurs at O_2 concentrations above 1% in gas phase. Remarkably, a reactor operating with O_2 levels below 0.5% air saturation efficiently cultivates microbes with varying oxygen requirements [102]. However, the restricted O_2 concentration below 1% stimulates GHP-N_2O production through the activity of NO_2^--N and NO reductases,

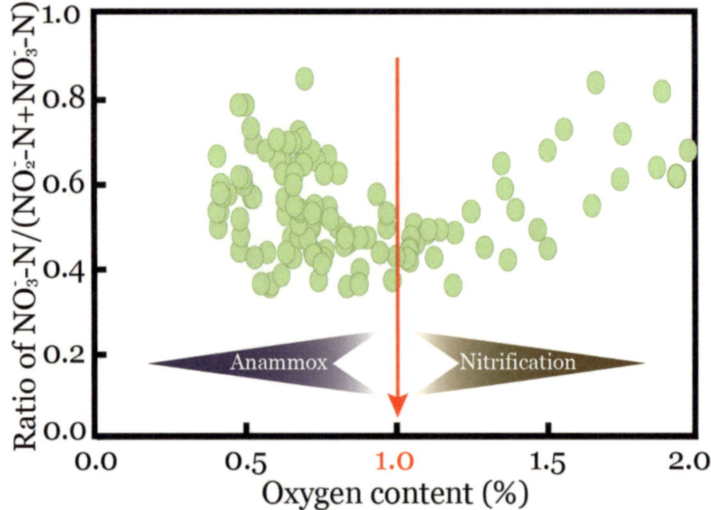

Figure 7.
Effect of O_2 in the downflow hanging sponge (DHS) for N removal on the production of NO_3^--N/ (NO_2^--N + NO_3^--N).

resulting in a loss of 7.2% nitrogen in the DHS. Considering the mass balance, it is observed that 55.7% of NH_4^+-N is utilized for NO_2^--N production, 34.5% is further transformed to gaseous N_2, but 9.5% is diverted toward the formation of N_2O under an O_2 concentration below 0.48% (**Figure 8**). In the DHS, a similar pattern of N_2O production was observed during nitritation, with N_2O accounting for 13% of the total nitrogen gas at an O_2 content of 0.4%. Additionally, the DHS supported the coexistence of six different nitrogen-functional microbes, namely aerobic ammonia-oxidizing *Nitrosomonas*, anaerobic ammonia-oxidizing *Brocadia*, canonical nitrite-oxidizing or comammox-functional *Nitrospira*, denitrifying *Comamonas* and nitrogen-fixing *Bradyrhizobium*. This diverse microbial community, facilitated

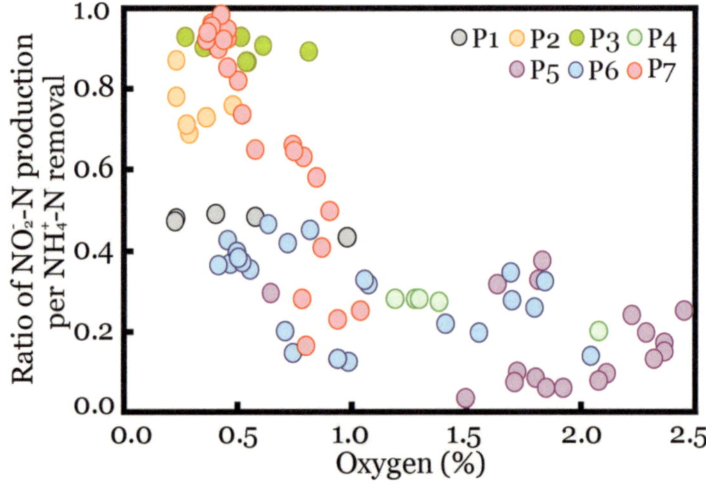

Figure 8.
Effect of O_2 in the downflow hanging sponge (DHS) for N removal on the ratio of N_2O-$N_{produced}$/NH_4^+-$N_{removed}$. Wherein, P1 ~ P7 is the data taken from phase 1 to phase 7 of the operation in the DHS.

by the DHS's excellent biomass retention capacity, created favorable conditions for the complete transformation of NH_4^+-N to N_2.

11. DHS-type systems applied for N_2O elimination

The emission of GHP-N_2O is commonly observed in various mixed systems involved in nitrogen transformation, particularly during NH_2OH oxidization [30], nitrifier denitrification [31], comammox [46] and NO_2^--N reduction [37]. Traditionally, the primary approach for N_2O removal is carried out through its reduction in the denitrification, which requires an adequate carbon source to activate the microbes possessing the N_2OR enzyme. However, this study explores an alternative possibility of N_2O oxidation catalyzed by inhered ammonia oxidizers, utilizing the abundant atmospheric O_2 as electron acceptor. The theoretically thermodynamic equations for these reactions are presented in Eqs. (1) and (2). These equations demonstrate the thermodynamic feasibility of N_2O oxidation to NO_3^--N by ammonia oxidizers using O_2 as the electron acceptor.

$$N_2O + O_2 + H_2O \rightarrow 2NO_2^- + 2H^+ \ \Delta G^0 = -5.33\,KJ/mole\,e^- \quad (1)$$

$$N_2O + 2O_2 + H_2O \rightarrow 2NO_3^- + 2H^+ \ \Delta G^0 = -21.2\,KJ/mole\,e^- \quad (2)$$

In order to assess the potential of N_2O oxidation, a series of experiments were conducted, including batch assay and continuous systems such as DHS and USS. The batch assay exhibited an average N_2O-removal rate of 107.1 mg N/g VSS (volatile suspended soilds) over a 60-day incubation period. During this time, NO_3^--N and N_2-N production accounted for 6.66 and 1.86% of the total nitrogen, respectively. The microbial community in the batch culture consisted of 93% domain *Bacteria* and 7% domain *Archaea*. Notably, the population of ammonia-oxidizing *Nitrososphaera gargensis*-like group experienced a 3.45-fold increase, representing 68.4% of the total archaeal population. Moving on to the continuous-flow DHS system, a range of N_2O-N loads, equivalent to HRT from 4 days to 12 hours, was applied under fully saturated O_2 conditions. The DHS demonstrated an impressive N_2O removal efficiency of 95% with a rate of 6.99 kg N/kg VSS-day. Concurrently, NO_3^--N production reached a maximum of 22.9 µmoles/day, while NH_4^+-N and NO_2^--N were not detected throughout the entire operation. These findings indicate the effective capability of the system in removing N_2O while promoting NO_3^--N production.

In the operation of the DHS, the concentration of NO_3^--N was accumulated in the effluent up to 7.79 mg N/L at the loadings below 1 mmole N_2O-N/day, whereas gaseous N_2 became the predominant end-product at the loadings over 2 mmoles N_2O-N/day [103]. The mass balance analysis, as illustrated in **Figure 9**, revealed that the produced N_2 was derived from two main pathways: the reduction of NO_3^- formed from the N_2O oxidation and the direct reduction of N_2O itself. The conversion yield of mole N_2-N per mole N_2O-N (0.109 as average of phases II-V) was approximately twice as high as the conversion yield of mole NO_3^--N per mole N_2O-N (0.052 in phase I). Furthermore, the anomalous increase in pH was observed in the final phase of the DHS operation, suggesting the potential accumulation of alkaline intermediates, such as NH_2OH. This observation lends support to the hypothesis that N_2O oxidation to NO_3^--N via NH_2OH as the intermediate may occur in the system.

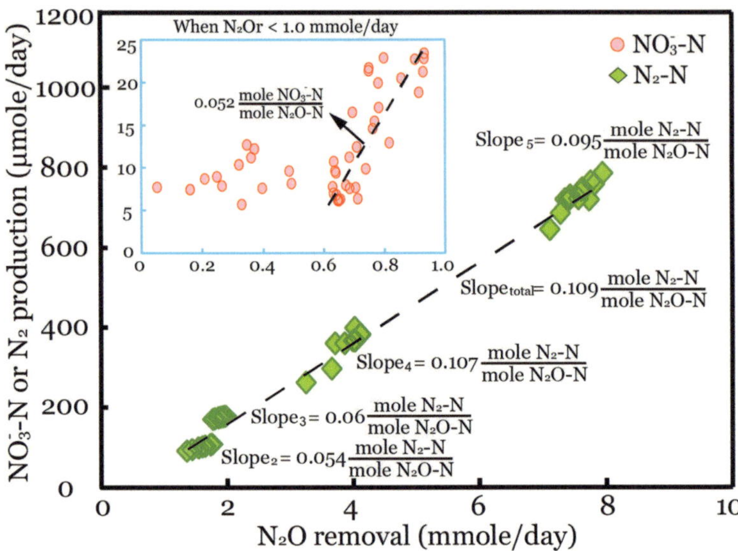

Figure 9.
The type of production from the transformed N_2O in the downflow hanging sponge (DHS).

In addition, a diverse community of the nitrogen-functional microbes coexisted in the DHS, including the dominant nitrite-oxidizing bacteria of the genus *Nitrospina* and ammonia-oxidizing archaea of the genus *Nitrosophaera* [104]. Furthermore, the family *Acidobacteriaceae* exhibited both denitrifying and DNRA activities [105]. A smaller population of ammonia-oxidizing bacteria was also detected (3.97 × 10^3 copies/μg DNA based on the *amoA* gene). The presence of the genera *Nitrosophaera* and *Nitrospira*, which are typically associated with marine environments, dominated in the aerobic DHS during N_2O transformation, potentially serving as regulators of marine N_2O production. Moreover, the contribution of nitrogen conversion through other redox reactions involving metals was estimated to be less than 10%. Therefore, we hypothesize that the major nitrogen loss observed in the DHS system could be attributed to the accumulation of unidentified intermediates resulting from N_2O transformation such as NH_2OH, nitrogen oxide (NO_x), dinitrogen trioxide (N_2O_3) and dinitrogen tetroxide (N_2O_4) and possibly other compounds that were not analyzed in this study.

Another system utilized for cultivating the N_2O-functional community is the submerged USS system, which received an average N_2O concentration of 10.96 ppm with N_2O load of 188 μmoles/day. Throughout the 240-day operation, the removal efficiency ranged from 13.6 to 37.5%, which resulted in effluent concentrations of NH_4^+-N and NO_2^--N ranging from 0.08 mgN/L to 0.60 mgN/L and 0.12 mgN/L to 0.44 mgN/L, respectively. After a 30-day enrichment, the production of NO_3^--N commenced, reaching a maximum concentration of 2.74 mgN/L, equivalent to a production rate of 47.0 μmoles/day, with 70.98% originating from the oxidation of N_2O on day 50, surpassing that of the aerobic DHS.

However, N_2 production began while the transformed NO_3^--N concentration was below 10%, and the maximum N_2 accumulation reached 53.6 μmoles/day, representing 5.4% of N_2O conversion. In contrast to the DHS system, where N_2 served as the end-product instead of NO_3^--N, the USS reactor exhibited N_2 production in conjunction with the presence of NO_3^--N. Additionally, the average of O_2 consumption during the 240-day operation ranged from 50.2 μmoles/day to 140.2 μmoles/day. Conversely, the concentrations of chloride and sulfate present in the cultivated medium experienced a reduction of over 78.2%, indicating their potential utilization by the nitrogen-functional microbes. Additionally, the activity of the N_2O oxidation, as determined by quantitative polymerase chain reaction (qPCR) analysis targeting functional genes, revealed a significant increase in the population of both ammonia oxidizers and nitrite oxidizers, reaching levels of approximately 10^5 cells/kg VSS. Notably, the growth rate of ammonia-oxidizing archaea outpaced that of ammonia-oxidizing bacteria, while the genus *Nitrospira* exhibited a higher growth rate compared to the genus *Nitrobacter*.

On the one hand, the four effect factors tested by the batch assay were discussed as follows. First, the influence of oxygen concentration was evaluated, and it was found that increasing the air input by 3-fold resulted in a 1.36-fold increase in NO_3^--N production compared to the aerobic DHS operated with 7% of air supply. This led to a NO_3^--N production rate of 10.16 mg/gVSS. Second, the addition of 34.5% methane (CH_4) led to the highest NO_3^--N production, reaching a value of 15.19 mg NO_3^--N/gVSS. Third, the enrichment of ammonia oxidizers using 7.5 mM NH_4^+-N in conjunction with N_2O-degrading microbes resulted in a conversion of 9.62 mg NO_3^--N/gVSS. Fourth, the addition of 1.4 mM manganese (Mn^{2+}) aimed to convert N_2O to NO_3^--N, but it achieved a lower conversion rate of 6.58 mg NO_3^--N/gVSS compared to the aerobic DHS. The microbial populations of both domains *Bacteria* and *Archaea* displayed an increase in the growth rate in response to O_2, CH_4 and NH_4^+-N, except in the case of the assay with Mn^{2+} addition. Notably, the increase in the population of nitrite oxidizing genera *Nitrobacter* and *Nitrospira* was superior than that of ammonia oxidizers.

To further confirm N_2O oxidation pathways involving NH_4^+-N oxidation and NO_2^--N oxidation, pure cultures of the genera *Nitrosomonas* and *Nitrobacter* were employed for N_2O elimination using sodium bicarbonate ($NaHCO_3$) and carbon dioxide (CO_2) as inorganic carbon sources. Ammonia-oxidizing *Nitrosomonas* demonstrated efficient catalysis of N_2O transformation, resulting in the production of 0.20 to 0.64 mole NO_x/mole N_2O and 0.30 to 0.54 mole NH_4^+-N/mole N_2O. The addition of a two fold carbon source facilitated an increase in N_2O oxidation. Notably, CO_3^{2-} derived from $NaHCO_3$ was more readily utilized for cell synthesis compared to gaseous CO_2. In contrast, the nitrite-oxidizing *Nitrobacter* exhibited lower N_2O transformation rates ranging from 25.94 to 53.84%. The addition of either inorganic carbon source resulted in high NO_x production from the oxidized N_2O, ranging from 0.90 to 0.91%. Based on these findings, it is suggested that the aerobic degradation of N_2O follows a possible route involving the conversion of N_2O to NO_3^--N via NH_2OH and NO_2^--N as the intermediate. Additionally, aerobic nitrogen-functional microbes were the key contributors for N_2O elimination.

12. Summary

Both NH_4^+-N and GHP-N_2O are of great concern due to their impact on the globally ecological environment. This chapter introduces the application of the polyurethane sponge as a useful medium, providing a three-dimensional space for microbial growth. Two types of sponge-based systems, namely DHS and USS reactors, were utilized for various reactions of autotrophic nitrogen transformation. In terms of NH_4^+-N conversion, processes, such as nitritation, anammox and one-stage nitrogen removal, demonstrated satisfactory rates of total nitrogen removal. Regarding N_2O elimination, three potential routes were identified for N_2O transformation, involving the production of NO_3^--N through the conversion of NO_2^--N, NH_4^+-N or direct conversion to N_2 as end-product. Five different autotrophic nitrogen-functional microbes cooperated synergistically within the expanded system, contributing to the reduction of nitrogenous compounds.

Acknowledgements

We are grateful to Professor Jer-Horng Wu who works in the Department of Environmental Engineering, National Cheng Kung University in Taiwan, for assisting with microbial clarification and data analysis of the N_2O oxidation system. We also thank 21 Century Center of Excellence (COE) program (Japan), Ministry of Science and Technology (Taiwan) and NCKU Research & Department Foundation (Taiwan) for supporting the funding.

Author details

Hui-Ping Chuang[1*], Akiyoshi Ohashi[2] and Hideki Harada[3,4]

1 Sustainable Environment Research Laboratories, National Cheng Kung University, Tainan, Taiwan

2 Department of Civil and Environmental Engineering, Hiroshima University, Hiroshima, Japan

3 DHS Technology, Inc., Tokyo, Japan

4 Department of Environmental Engineering, Tohoku University, Sendai, Japan

*Address all correspondence to: hpchuang@mail.ncku.edu.tw

References

[1] NOAA: Global Monitoring Laboratory. Carbon cycle greenhouse gases [Internet]. 2022. Available from: https://gml.noaa.gov/ccgg/trends/

[2] Ravishankara AR, Daniel JS, Portmann RW. Nitrous oxide (N_2O): The dominant ozone-depleting substance emitted in the 21st century. Science. 2009;**326**:123-125

[3] Kuypers MMM, Marchant HK, Kartal B. The microbial nitrogen-cycling network. Nature Reviews Microbiology. 2018;**16**:263-276

[4] Gupta VK, Sadegh H, Yari M, Ghoshekandi RS, Maazinejad B, et al. Removal of ammonium ions from wastewater-a short review in development of efficient methods. Global Journal of Environmental Science and Management. 2015;**1**(2):148-158

[5] Ming T, de Richter R, Chen S, Caillol S. Fighting global warming by greenhouse gas removal destroying atmospheric nitrous oxide thanks to synergies between two breakthrough technologies. Environmental Science and Pollution Research. 2016;**23**:6119-6138

[6] Monfet E, Aubry G, Ramirez AA. Nutrient removal and recovery from digestate: A review of the technology. Biofuels. 2018;**9**(2):247-262

[7] Gruber N, Galloway JN. An earth-system perspective of the global nitrogen cycle. Nature. 2008;**451**:293

[8] IPCC. Climate Change: The Physical Science Basis. Contribution of Working Group I to the Fifth Assessment Report of the Intergovernmental Panel on Climate Change. Cambridge University Press; 2013. DOI: 10.1017/CBO9781107415324

[9] US EPA: Greenhouse Gas Emissions [Internet]. 2020 Available from: https://www.epa.gov/ghgemissions/overview-greenhouse-gases

[10] Pérez-Ramírez J, Kapteijn F, Schöffel K, Moulijn JA. Formation and control of N_2O in nitric acid production: Where do we stand today? Applied Catalysis B: Environmental. 2003;**44**:117-151

[11] Kong Q, Zhang J, Miao M, Tian L, Guo N, Liang S. Partial nitrification and nitrous oxide emission in an intermittently aerated sequencing batch biofilm reactor. Chemical Engineering Journal. 2013;**217**:435-441

[12] Yang JJ, Trela J, Plaza E, Tjus K. N_2O emissions from a one stage partial nitrification/anammox process in moving bed biofilm reactors. Water Science & Technology. 2013;**68**:144-152

[13] Eldyasti A, Nakhla AG, Zhu J. Influence of biofilm thickness on nitrous oxide (N_2O) emissions from denitrifying fluidized bed bioreactors (DFBBRs). Journal of Biotechnology. 2014;**192** (Part A):281-290

[14] Tode D, Dorsch P. Nitrous oxide emissions in a biofilm loaded with different mixtures of concentrated household wastewater. International Journal of Environmental Science and Technology. 2015;**12**:1-12

[15] Red Y, Ngo HH, Guo W, Ni B-J, Liu Y. Linking the nitrous oxide production and mitigation with the microbial community in wastewater treatment: A review. Bioresource Technology Reports. 2019;**7**:100191

[16] Frutos OD. Novel Biotechnologies for Nitrous Oxide Abatement [thesis]. Valladolid: University of Valladolid; 2018

[17] Prosser JI, Hink L, Gubry-Rangin C, Nicol GW. Nitrous oxide produced by ammonia oxidizers: Physiological diversity, niche differentiation and potential mitigation. Global Change Biology. 2020;**26**:103-118

[18] Centi G, Perathoner S, Rak ZS. Reduction of greenhouse gas emissions by catalytic processes. Applied Catalysis B: Environmental. 2003;**41**:143-155

[19] Dell R, Stone F, Tiley P. The decomposition of nitrous oxide on cuprous oxide and other oxide catalysts. Transactions of the Faraday Society. 1953;**49**:201-209

[20] Blyholder G, Tanaka K. Photocatalytic reactions on semiconductor surfaces: I. decomposition of nitrous oxide on zinc oxide. The Journal of Physical Chemistry. 1971;**75**:1037-1043

[21] McCarty PL. What is the best biological process for nitrogen removal: When and why? Environmental Science & Technology. 2018;**52**:3835-3841

[22] Hanaki K, Hong Z, Matsuo T. Production of nitrous oxide gas during denitrification of wastewater. Water Science & Technology. 1992;**26**(5-6):1027-1036

[23] Mania D, Heylen K, van Spanning RJM, Frostegard Å. Regulation of nitrogen metabolism in the nitrate-ammonifying soil bacterium *bacillus vireti* and evidence for its ability to grow using N_2O as electron acceptor. Environmental Microbiology. 2016;**18**(9):2937-2950

[24] Chuang H-P, Wu J-H, Ohashi A, Abe K, Hatamoto M. Potential of nitrous oxide conversion in batch and down-flow hanging sponge bioreactor systems. Sustainable Environment Research. 2014;**24**(5):117-128

[25] Brock TD, Madigan MT. Martinko JM and J Parker: Biology of Microorganisms. 8th ed. Upper Saddle River NJ, USA: Prebtice Hall; 1997

[26] Koops HP, Pommerening-Röser A. Distribution and ecophysiology of the nitrifying bacteria emphasizing cultured species. FEMS Microbiology Ecology. 2001;**37**:1-9

[27] Cooper AB. Nitrate depletion in the riparian zone and stream channel of a small headwater catchment. Hydrobiologia. 1990;**202**:12-36

[28] Zumft WG. The denitrifying prokaryotes. In: Balows A, Truper HG, Dworkin M, Harder W, Scheleifer KH, editors. 2nd Ed., the Prokaryote a Handbook on the Biology of Bacteria: Ecophysiology Isolation Identification Applications. 2nd ed. Vol. 1. New York, NY: Springer-Verlag; 1992. pp. 554-582

[29] Hatamoto M, Okubo T, Kubota K, Yamaguchi T. Characterization of downflow hanging sponge reactors with regard to structure, process function and microbial community composition. Applied Microbiology and Biotechnology. 2018;**102**:10345-10352

[30] Goreau TJ, Kaplan WA, Wofsy SC, McEloy MB, Valois FW, Watson SW. Production of NO_2^- and N_2O by nitrifying bacteria at reduced concentrations of oxygen. Applied and Environmental Microbiology. 1980;**40**(3):526-532

[31] Wrage N, Velthof GL, van Beusichem ML, Oenema O. Role of

nitrifier denitrification in the production of nitrous oxide. Soil Biology and Biochemistry. 2001;**33**(12):1723-1732

[32] Wunderlin P, Lehmann MF, Siegrist H, Tuzson B, Joss A, Emmenegger L, et al. Isotope signatures of N_2O in a mixed microbial population system: Constraints on N_2O producing pathways in wastewater treatment. Environmental Science & Technology. 2013;**47**:1339-1348

[33] Law Y, Ni BJ, Lant P, Yuan ZG. N_2O production rate of an enriched ammonia-oxidising bacteria culture exponentially correlates to its ammonia oxidation rate. Water Research. 2012;**46**:3406-3419

[34] Kostera J, Mcgarry J, Pacheco AA. Enzymatic interconversion of ammonia and nitrite: The right tool for the job. Biochemistry. 2010;**49**:8546-8553

[35] Schmidt I, Bock E. Anaerobic ammonia oxidation with nitrogen dioxide by *Nitrosomonas eutropha*. Archives of Microbiology. 1997;**167**:106-111

[36] Hooper AB. A nitrite-reducing enzyme from *Nitrosomonas europaea* - preliminary characterization with hydroxylamine as electron donor. Biochimica et Biophysica Acta. 1968;**162**:49

[37] Beaumont HJE, van Schooten B, Lens SI, Westerhoff HV, van Spanning RJM. *Nitrosomonas europaea* expresses a nitric oxide reductase during nitrification. Journal of Bacteriology. 2004;**186**:4417-4421

[38] Kostera J, Youngblut MD, Slosarczyk JM, Pacheco AA. Kinetic and product distribution analysis of NO center dot reductase activity in *Nitrosomonas europaea* hydroxylamine oxidoreductase. Journal of Biological Inorganic Chemistry. 2008;**13**:1073-1083

[39] Beaumont HJE, Hommes NG, Sayavedra-Soto LA, Arp DJ, Arciero DM, et al. Nitrite reductase of *Nitrosomonas europaea* is not essential for production of gaseous nitrogen oxides and confers tolerance to nitrite. Journal of Bacteriology. 2002;**184**:2557-2560

[40] Bergaust L, van Spanning RJM, Frostegard A, Bakken LR. Expression of nitrous oxide reductase in *Paracoccous denitrificans* is regulated by oxygen and nitric oxide through FnrP and NNR. Microbiology. 2012;**158**:826-834

[41] Pan Y, Ni B-J, Yuan Z. Modeling electron competition among nitrogen oxides reduction and N_2O accumulation in denitrification. Environmental Science & Technology. 2013;**47**:11083-11091

[42] Campos JL, Arrojo B, Vazquez-Padín JR, Mosquera-Corral A, Méndez R. N_2O production by nitrifying biomass under anoxic and aerobic conditions. Applied Biochemistry and Biotechnology. 2009;**152**(2):189-198

[43] Fernandes AT, Damas JM, Todorovic S, Huber R, Baratto MC, et al. The multicopper oxidase from the archaeon *Pyrobaculum aerophilum* shows nitrous oxide reductase activity. The FEBS Journal. 2010;**277**(15):3176-3189

[44] Cross R, Lloyd D, Poole RK, Moir JWB. Enzymatic removal of nitric oxide catalyzed by cytochrome c' in *Rhodobacter capsulates*. Journal of Bacteriology. 2001;**183**(10):3050-3054

[45] Kern M, Simon J. Electron transport chains and bioenergyetics of respiratory nitrogen metabolism in *Wolinella succinogenes* and other *Epsilonproteobacteria*. Biochimica et Biophysica Acta-Bioenergetics. 2009;**1787**(6):646-656

[46] Pinto A, Marcus DN, Ijaz UZ, Bautista-de Lose Santos QM, et al.

Metagenomic evidence for the presence of comammox *Nitrospira*-like *bacteria* in a drinking water system. Applied Environmental Science. 2015;**1**(1):e00054-e00015

[47] Jetten MSM, Wagner M, Furest J, van Loosdrecht MCM, Kuenen JG, et al. Microbiology and application of the anaerobic ammonium oxidation (anammox) process. Current Opinion in Biotechnology. 2001;**12**:283-288

[48] Hellinga C, Schellen AAJC, Mulder JW, van Loosdrecht MCM, et al. The SHARON process: An innovative method for nitrogen removal from ammonium-rich wastewater. Water Science & Technology. 1998;**37**:135-142

[49] Schmidt I, Sliekers O, Schmid M, Bock E, Fuerst J, et al. New concepts of microbial treatment processes for the nitrogen removal in wastewater. FEMS Microbiology Reviews. 2003;**27**:481-492

[50] Sliekers AO, Derwort N, Campos Gomez JL, Strous M, Kuenen JG, Jetten MSM. Complete autotrophic nitrogen removal over nitrite in one single reactor. Water Research. 2022;**36**:2475-2482

[51] Haroon MF, Hu S, Shi Y, Imelfort M, Keller J, et al. Anaerobic oxidation of methane coupled to nitrate reduction in a novel archaeal lineage. Nature. 2013;**500**:567-570

[52] Kuai L, Verstraete W. Ammonium removal by the oxygen-limited autotrophic nitrification-denitrification system. Applied and Environmental Microbiology. 1998;**64**:4500-4506

[53] Van Kessel MAHJ, Stultiens K, Slegers MFW, Cruz SG, Jetten MSM, et al. Current perspectives on the application of N-damo and anammox in wastewater treatment. Current Opinion in Biotechnology. 2018;**50**:222-227

[54] Hippen A, Rosenwinkel KH, Baumgerten G, Seyfriend CF. Aerobic deammonification - a new experience in the treatment of wastewaters. Water Science & Technology. 1997;**35**:111

[55] Caranto JD, Lancaster KM. Nitric oxide is an obligate bacterial nitrification intermediate produced by hydroxylamine oxidoreductase. Proceedings of the National Academy of Sciences of the United States of America. 2017;**114**:8217-8222

[56] Liu S et al. Abiotic conversion of extracellular NH_2OH contributes to N_2O emission during ammonia oxidation. Environmental Science & Technology. 2017;**51**:13122-13132

[57] Bakken LR, Bergaust L, Liu B, Frostegård Å. Regulation of denitrification at the cellular level: A clue to the understanding of N_2O emissions from soils. Philosophical Transactions of the Royal Society B. 2012;**367**:1226-1234

[58] Hayatsu M, Tago K, Uchiyama I, Toyoda A, Wang Y, et al. An acid-tolerant ammonia-oxidizing gamma-*Proteobacterium* from soil. The ISME Journal. 2017;**11**:1130-1141

[59] De Bruijn P, van de Graaf AA, Jetten MSM, Robertson LA, Kuenen JG. Growth of *Nitrosomonas europaea* on hydroxylamine. FEMS Microbiology Letters. 1995;**125**:179-184

[60] Lehtovirta-Morley LE, Stoecker K, Vilcinskas A, Prosser JI, Nicol GW. Cultivation of an obligate acidophilic ammonia oxidizer from a nitrifying acid soil. PNAS. 2011;**108**:15892-15897

[61] Lebedeva EV, Hatzenpichler R, Pelletier E, Schuster N, Hauzmayer S, et al. Enrichment and genome sequence of the group I.1a ammonia-oxidizing archaeon "Ca. *Nitrosotenuis uzonensis*"

representing a clade globally distributed in thermal habitats. PLoS One. 2013;**8**(11):e80835

[62] Lehtovirta-Morley LE, Ross J, Hink L, Weber EB, Gubry-Rangin C, et al. Isolation of Candidatus *Nitrosocomicus franklandus*, a novel ureolytic soil archaeal ammonia oxidiser with tolerance to high ammonia concentration. FEMS Microbiology Ecology. 2016;**92**:10

[63] Ahlgren NA, Chen Y, Needham DM, Parada AE, Sachdeva R, et al. Genome and epigenome of a novel marine *Thaumarchaeota* strain suggest viral infection, phosphorothioation DNA modification and multiple restriction systems. Environmental Microbiology. 2017;**19**(6):2434-2452

[64] Kartal B, van Niftrik L, Keltjens JT, den Camp H, Jetten MSM. Anammox-growth physiology, cell biology, and metabolism. In: Poole RK, editor. Advances in Microbial Physiology. Vol. 60. London, UK: Academic Press and Elsevier Science; 2012. pp. 211-262

[65] Kuenen JG. Anammox bacteria: From discovery to application. Nature Reviews. Microbiology. 2008;**6**:320-326

[66] Zhao R, Biddle JF, Jorgensen SL. Introducing Candidatus *Bathyanammoxibaceae*, a family of bacteria with the anammox potential present in both marine and terrestrial environments. ISME Communications. 2022;**2**:42

[67] Jetten MSM, Camp HJM, Kuenen JG, Strous M. Order II. "*Candidatus Brocadiales*" ord. nov.," In: Krieg NR, Ludwig W, Whitman WB, Hedlund BP, Paster BJ, et al., editors. The *Bacteroidetes*, Spirochaetes, *Tenericutes* (*Mollicutes*), *Acidobacteria, Fibrobacteres, Fusobacteria, Dictyoglomi, Gemmatimonadetes, Lentisphaerae, Verrucomicrobia, Chlamydiae*, and *Planctomycetes*. 2nd Edn. New York: Springer; 2010. 918-925

[68] Kartal B, Rattray J, van Niftrik LA, van de Vossenberg J, Schmid MC, et al. Candidatus "*Anammoxoglobus propionicus*" a new propionate oxidizing species of anaerobic ammonium oxidizering bacteria. Systematic and Applied Microbiology. 2007;**30**(1):39-49

[69] Kartal B, Kuypers MMM, Lavik G, Schalk J, et al. Anammox bacteria disguised as denitrifiers: Nitrate reduction to dinitrogen gas via nitrite and ammonium. Environmental Microbiology. 2007;**9**(3):635-642

[70] Daims H, Lebedeva EV, Pjevac P, Han P, Herbold C, et al. Complete nitrification by *Nitrospira* bacteria. Nature. 2015;**528**:24-31

[71] Navada S, Vadstein O, Tveten A-K, Verstege GC, Terjesen BF, Mota VC, et al. Influence of rate of salinity increase on nitrifying biofilm. Journal of Cleaner Production. 2019;**238**:117835

[72] Koch H, van Kessel MAHJ, Lucker S. Complete nitrification: Insights into the ecophysiology of comammox Nitrospira. Applied Environmental Biotechnology. 2019;**103**:177-189

[73] Kits KD et al. Kinetic analysis of a complete nitrifier reveals an oligotrophic lifestyle. Nature. 2017;**549**:269-272

[74] Shao M-F, Zhang T, Fang HHP. Sulfur-driven autotrophic denitrification: Diversity, biochemistry and engineering applications. Applied Microbiology and Biotechnology. 2010;**88**:1027-1042

[75] Domeignoz-Horta L et al. Non-denitrifying nitrous oxide-reducing bacteria - an effective N. Soil Biology and Biochemistry. 2016;**103**:376-379

[76] Franche C, Lindstrom K, Elmerich C. Nitrogen-fixing bacteria associated with leguminous and non-leguminous plants. Plant and Soil. 2009;**321**:35-39

[77] Kondaveeti S, Abu-Reesh IM, Mohanakrishna G, Bulut M, Pant D. Advanced routes of biological and bio-electrocatalytic carbon dioxide (CO2) mitigation toward carbon neutrality. Frontiers in Energy Research. 2020;**8**:94

[78] Könneke M, Schubert DM, Brown PC, Hügler M, Standfest S, et al. Ammonia-oxidizing archaea use the most energy-efficient aerobic pathway for CO_2 fixation. PNAS. 2014;**111**(22):8239-8244

[79] Badger MR, Bek EJ. Multiple rubisco forms in proteobacteria: Their functional significance in relation to CO_2 acquisition by the CBB cycle. Journal of Experimental Botany. 2008;**59**(7):1525-1541

[80] Alfreider A, Grimus V, Luger M, Ekblad A, Salcher M, Summerer M. Autotrophic carbon fixation strategies used by nitrifying prokaryotes in freshwater lakes. FEMS Microbiology Ecology. 2018;**94**:fry163

[81] Bayer B, McBeain K, Carlson CA, Santoro AE. Carbon content, carbon fixation yield and dissolved organic carbon release from diverse marine nitrifiers. Limnology and Oceanography. 2023;**68**:84-96

[82] Chuang HP. Development of a Novel and Cost-Effective Nitrogen Removal Process by the Application of Down-Flow Hanging Sponge (DHS) Systems [thesis]. Niigata: Nagaoka University of Technology; 2007

[83] Pynaert K, Smets BF, Beheydt D, Verstraete W. Start-up of autrophic nitrogen removal reactors via sequential biocatalyst addition. Environmental Science & Technology. 2004;**38**:1228-1235

[84] Gao M, Yang M, Li H, Yang Q, Zhang Y. Comparison between a submerged membrane bioreactor and a conventional activated sludge system on treating ammonia-bearing inorganic wastewater. Biotechnology. 2004;**108**:265-269

[85] Chuang HP, Ohashi A, Imachi H, Tandukar M, Harada H. Effective partial nitrification to nitrite by down-flow hanging sponge reactor under limited oxygen condition. Water Research. 2007;**41**(2):295-302

[86] Wyffels S, Boeckx P, Pynaert K, Zhang D, et al. Nitrogen removal from sludge reject water by a two-stage oxygen-limited autotrophic nitrification denitrification process. Water Science & Technology. 2004;**49**(5-6):57-64

[87] Park H-D, Noguera DR. Evaluating the effect of dissolved oxygen on ammonia-oxidizing bacterial communities in activated sludge. Water Research. 2004;**38**(14-15):3275-3286

[88] Tokutomi T. Operation of a nitrite-type airlift reactor at low DO concentration. Water Science & Technology. 2004;**49**(5-6):81-88

[89] Bernet N, Dangjong N, Delgenes JP, Moletta R. Nitrification at low oxygen concentration in biofilm reactor. Journal of Environmental Engineering. 2001;**127**(3):266-271

[90] Chuang HP, Imachi H, Tandukar M, Kawakami S, Harada H, Ohashi A. Microbial community that catalyzes partial nitrification at low oxygen atmosphere as revealed by 16S rRNA and *amoA* genes. Journal of Bioscience and Bioengineering. 2007;**104**(6):525-528

[91] Strous M, Heijnen JJ, et al. The sequencing batch reactor as a powerful tool for the study of slowly growing

anaerobic ammonium-oxidizing microorganisms. Applied Microbiology and Biotechnology. 1998;**50**:589-596

[92] Furukawa K, Tokutomi T, Imajo U. Granulation of anammox microorganisms in up-flow reactors. Water Science & Technology. 2004;**49**(5-6):155-163

[93] Tsushima I, Kindaichi T, Okabe S. Quantification of anaerobic ammonium-oxidizing bacteria in enrichment cultures by real-time PCR. Water Research. 2007;**41**:785-794

[94] Chuang HP, Yamaguchi T, Harada H, Ohashi A. Anoxic ammonium oxidation by application of a down-flow hanging sponge (DHS) reactor. Journal of Environmental Engineering. 2008;**18**(6):409-417

[95] Strous M et al. Missing lithotroph identified as new *Planctomycete*. Nature. 1999;**400**:446-449

[96] Moreno-Vivián C, Cabello P, Martínez-Luque M, et al. Prokaryotic nitrate reduction: Molecular properties and functional distinction among bacterial nitrate reductases. Journal of Bacteriology. 1999;**181**(21):6573-6584

[97] Sliekers AO, Third KA, Abma W, Kuenen JG, Jetten MSM. CANON and Anammox in a gas-lift reactor. FEMS Microbiology Letters. 2003;**218**: 339-344

[98] Fux C, Siegrist H. Nitrogen removal from sludge digester liquids by nitrification/denitrification or partial nitrification/anammox: Environmental and economical considerations. Water Science & Technology. 2004;**50**(10):19-26

[99] Wen A, Fegan M, Hayward C, Chakraborty S, Sly LI. Phylogenetic relationships among members of the *Comamonadaceae*, and description of *Delftia acidovorans* (den Dooren de Jong 1926 and Tamaoka et al., 1987) gen. Nov., comb. Nov. International Journal of Systematic Bacteriology. 1999;**49**:567-576

[100] Khan ST, Hiraishi A. *Diaphorobacter nitroreducens* gen. Nov., sp., nov., a poly(3-hydroxybuturate)-degrading denitrifying bacterium isolated from activated sludge. The Journal of General and Applied Microbiology. 2002;**48**:299-308

[101] Third KA, Sliekers AO, Kuenen JG, Jetten MSM. The CANON system (completely autotrophic nitrogen-removal over nitrite) under ammonium limitation: Interaction and competition between three groups of bacteria. System. Applied Microbiology. 2001;**68**:1312-1318

[102] Shivaraman N, Shivaraman G. Anammox - a novel microbial process for ammonium removal. Current Science India. 2003;**84**(12):1507-1508

[103] Chuang H-P, Wu J-H, Ohashi A. Abe and Hatamoto M: Potential of nitrous oxide conversion in batch and down-flow hanging sponge bioreactor systems. Sustainable Environment Research. 2014;**24**(5):117-128

[104] De La Torre JR, Walker CB, Ingalls AE, Kolnneke M, Srahl DA. Cultivation of a thermophilic ammonia oxidizing archaeon synthesizing *Crenarchaeol*. Environmental Microbiology. 2008;**10**:810-818

[105] Eichorst SA, Kuske CR, Schmidt TM. Influentce of plant polymers on the distribution and cultivation of bacteria on the phylum *Acidobacteria*. Applied and Environmental Microbiology. 2011;77(2):586-596

Chapter 5

Removal of Nitrate and Nitrite by Donnan Dialysis: Optimization According to Doehlert Design

Ikhlass Marzouk Trifi, Beyram Trifi and Lasâad Dammak

Abstract

The removal of nitrate and nitrite simultaneously was investigated by Donnan dialysis (DD) using a Response Surface Methodology (RSM) approach. DD is a membrane process that consists of cross-ion exchange having the same electric charge through an ion-exchange membrane separating two solutions. In addition to being easy to handle, DD process is continuous, economical, requiring only few chemicals and low pumping energy. Statistical tools were applied to investigate the simultaneous removal of nitrates and nitrites by DD. The RSM is an efficient statistical strategy to design experiments, build models, determine the optimum conditions, and evaluate the significance of factors, even the interaction between them. A preliminary study was performed with three commercial membranes (AFN, AMX, ACS) in order to determine the experimental field based on different parameters. Then, a full-factor design was developed to determine the influence of these parameters and their interactions on the removal of nitrates and nitrites by DD. The RSM was applied according to the Doehlert model to determine the optimum conditions. The use of the RSM can be considered a good solution to determine the optimum condition compared to the traditional "one-at-a-time" method.

Keywords: nitrate, nitrite, Donnan dialysis, anion-exchange membrane, Doehlert

1. Introduction

The growth of population, industrialization, and rapid urbanization increases the pollution of nitrogen (N). Nitrate and nitrite are often used as pollutants; they present a water-quality problem [1]. Nitrate is primarily responsible for water eutrophication and infectious diseases in the environment [2]. In comparison to nitrate, the nitrite has a higher level of toxicity to human health. In the body, methemoglobin can be formed when it combines with hemoglobin, reducing oxygen transport. Additionally, this compound can be converted into carcinogenic nitrosamine, which is linked to hypertension, leukemia, brain tumors, stomach cancers, and bowel cancers [3]. The World Health Organization (WHO) suggests a maximum nitrate content of 50 mg/L in irrigation water and 0.5 mg/L in drinking water to protect the public health from the harmful effects of high nitrate and nitrite concentrations [4].

It was possible to remove nitrate and nitrite chemically using chlorination and physicochemically using coagulation-flocculation [5]. In contrast to traditional biological approaches [6], bioadsorption [7], and photocatalytic denitrification [8], most methods require oxidation in order to remove nitrite. Removing nitrate and nitrite simultaneously avoids this oxidation step, which requires chemical products to turn nitrite into nitrate.

Donnan dialysis (DD) was chosen because it is mostly economical. Because it only needs, a small amount of chemicals, pumping energy, and is simple to operate; this process is continuous and inexpensive. In the DD process, an ion-exchange membrane separates a compartment containing the solution to be treated (feed) from a compartment containing the solution that receives the target ions (receiver). The concentration gradients of the ions transported through the membrane (counter-ions) control the ion-exchange kinetics [9, 10]. The DD process involves the stoichiometric exchange of counter-ions, or ions with the same charge, over an ion-exchange membrane, and it is ended when Donnan equilibrium is attained [11].

As is common knowledge, the DD process is used to purify, concentrate, and remove various ions from wastewater and industrial effluents, including boron [12, 13], fluoride [14, 15], chromium [16, 17] and nitrates, nitrites [18, 19] using different type of anion-exchange membranes. Although Donnan dialysis has certain advantages in terms of cost and energy efficiency, industry does not adopt it primarily due to its slow kinetics. The almost of applications have been studied at laboratory scale.

The typical "one factor at a time" method of optimizing multivariate systems is not only time-consuming, but also often does not take into account the effects of cross-interactions between experimental factors. Furthermore, this approach implies that the best levels must be determined by multiplying experiments, which is not always true. By integrating Doehlert's experimental design [20] with response surface methods to simultaneously optimize all influential parameters, these drawbacks of the single-factor optimization procedure can be avoided.

The Box and Wilson-developed Response Surface Methodology (RSM) is a set of mathematical and statistical methods for studying situations like the one that is being posed using an empirical model. The RSM is a useful tool for process optimization. The benefit of this method over the traditional one is that it takes less time and costs less. Doehlert designs have a number of advantages over other designs, such as central composite or Box-Behnken designs. Because the number of levels can vary from one variable to another, there is greater flexibility in assigning a high or low number of levels to the selected variables, saving time on studies. Additionally, adjacent hexagons may successfully occupy a space because they do not overlap, making Doehlert designs more effective at mapping space [20].

The present investigation presents the application of the RSM applying Doehlert experimental design studies to look into the simultaneous removal of nitrates and nitrites by Donnan dialysis. In order to set up the experimental field, one component (nitrate or nitrite) was first eliminated from the feed compartment using different parameters, such as the concentration of counter-ions and the concentration of nitrate and nitrite separately. In order to improve the procedure and comprehend the simultaneous transport of nitrites and nitrites, the removal of two components (nitrate and nitrite) in the feed compartment was then explored by the Response Surface Methodology (RSM) by Doehlert design.

2. Experimental

2.1 Anionic-exchange membranes

During the Donnan dialysis operation, three commercial anion-exchange membranes have been used: ACS, AMX, and AFN. These membranes have the same structural properties and are homogeneous. **Table 1** presents the characteristics and properties of these membranes.

In order to prepare the samples for use in the Donnan dialysis, they had to be conditioned before any measurement. This was done primarily to eliminate contaminants from the production process and to stabilize their physical-chemical properties. French standard NF X 45-200 [21] was followed in doing this conditioning.

2.2 Donnan dialysis

All Donnan dialysis experiments were carried out using a laboratory cell. The device was used to study the nitrate and nitrite removal by Donnan dialysis. It is composed of a thermoregulated water bath (25.0 ± 0.1°C in this study), containing a cell with feed and receiver compartments separated by an anion-exchange membrane.

The dialysis cell consists of two detachable compartments made with polymethylmetacrylate (plexiglass) as shown in **Figure 1a** and **b**, in two formats, a photo of the mounted cell ready to be connected to the peristaltic pump, and a drawing of the

Membrane	Ion-exchange capacity (mmol/g)	Water content %	Thickness (mm)
AFN	3.00	47.8	0.12
AMX	1.30	26.0	0.13
ACS	1.85	18.9	0.15

Table 1.
The main characteristics of the three commercial anion-exchange membranes.

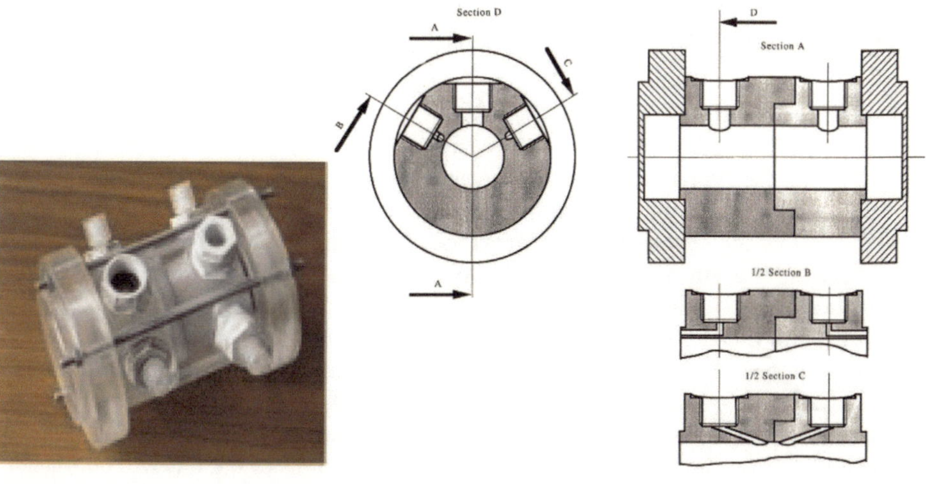

Figure 1.
The two-compartment cell used for the Donnan dialysis experiments. (a) Photo of an assembled cell. (b) Plan with different cuts according to three sections.

different sections of this cell. It is composed of four parts joined by three stainless steel threaded rods. The centering is assured by bolsters. The two central compartments, consisting of two tubes are symmetrical. Two threaded holes penetrate each compartment and serve as support for stuffing boxes. The membrane is sandwiched between these two compartments, making a seal at the same time [22].

Then, the solution of nitrate or/and nitrite was prepared as a feed compartment with different concentrations varying from 50 to 500 mg/L, and the solution of chloride with different concentrations varying from 10 to 100 mg/L. The solutions are placed in volumetric flasks which are essential due to their geometry because they limit the losses of solvent by evaporation. There the circulation of these solutions in the two compartments is ensured by a peristaltic pump equipped with two identified heads controlled by a speed variator acting on his engine. The circulation of fluids takes place in flexible reference pipes. The volume flow rates Q_f and Q_R of solutions leaving the feed and receiver compartments respectively are determined using a 1000 mL flask and a stopwatch measuring their filling times. This method assumes that the transmembrane volume flux is infinitely small compared to the solution flow rate imposed in each compartment.

The residual concentration of nitrate or/and nitrite at the outlet of the receiver compartment was determined spectrophotometrically. The UV-spectrophotometry approach was used to measure the nitrite and nitrate content at the receiver compartment during the dialysis operations [23]. The sodium salicylate and nitrate reaction produce paranitrosalicylate sodium, which is yellow colored. This reaction is followed by absorbance measurements at 415 nm using a UV-visible spectrophotometer to determine the amount of nitrate present. The amino-4-benzenesulfonamide was diazotized by nitrites in an acidic medium, and when it was coupled with N-(naphthyl-1) diamino-1,2-ethane dichloride, a purple-colored complex resulted. This complex's absorbance at 543 nm was then measured using a UV-visible spectrophotometer. The removal rate of nitrate (Y_1%) and nitrite (Y_2%) was calculated by Eq. (1):

$$Y_{1\text{ or }2}(\%) = \frac{C_0 - C_e}{C_0} \times 100 \qquad (1)$$

where C_e is the nitrate and nitrite equilibrium concentration (mg/L) and C_0 is the initial concentration of nitrate or nitrite (mg/L).

2.3 Doehlert design

A Doehlert design based on the RSM was employed as the experimental strategy in this inquiry. The values of these parameters must be simultaneously optimized in order to achieve the best system performance. The optimal scenario was found by superimposing the contours of the response surfaces in a plot with multiple responses. In the three-dimensional plots of numerous variables used to represent the graphical optimization in the experimental field, the regions of optimal response would be highlighted in red. A close match between the experimental and projected values is required [24]. The total number of experiments for k factors is $N = k^2 + k + 1$. In fifteen tests, three duplicates at center field were employed [24–27].

The initial nitrate, nitrite, and counter-ion concentrations in the receiver compartment were the factors that were examined. With one component in the feed compartment, the range of these factors was set in accordance with the preliminary investigation. The experimental field of the factors under investigation is shown in **Table 2**.

Factors	Symbol	Range and levels		
Coded variable X_1	$[NO_2^-]$	−1	0	1
Initial concentration NO_2^- (mg/L)		10	55	100
Coded variable X_2	$[NO_3^-]$	−1	0	1
Initial concentration NO_3^- (mg/L)		50	275	500
Coded Variable X_3	$[Cl^-]$	−1	0	1
Concentration of Cl^- (mol/L)		0.1	0.3	0.5

Table 2.
Range and levels of nitrate and nitrite removal.

The variables that were examined included the initial nitrate, nitrite, and counter-ion concentrations in the receiver compartment. The range of these factors was established with one component in the feed compartment in accordance with the preliminary investigation. The experimental setting for the factors under investigation is shown in **Table 2**.

The Doehlert design, a matrix that can anticipate the values of the response at any point in the experimental area, can be used to estimate the coefficients of a second-order function [28]. Using a polynomial equation (Eq. (2)), the selected model represents the predicted values of the answers Y. Bi represents the estimated major effect of component i, b_{ii} represents the estimated second-order effects, b_{ij} represents the estimated interactions between the factors i and j, and X_i represents the coded variable. NemrodW® Software was used to determine the model's coefficients.

$$Y_{1 \text{ or } 2} = b_0 + b_1X_1 + b_2X_2 + b_3X_3 + b_{11}X_1^2 + b_{22}X_2^2 + b_{33}X_3^2 + b_{12}X_1X_2 + b_{13}X_1X_3 + b_{23}X_2X_3 \quad (2)$$

The percentage absolute error of deviation (AED) and the regression coefficient (R^2) between experimental and theoretical findings were employed as two metrics to assess the models. Eq. (3) was used to compute the AED.

$$\text{AED } (\%) = \frac{100}{N} \cdot \left| \frac{Y_{exp} - Y_{theo}}{Y_{exp}} \right| \quad (3)$$

where Y_{theo} represents the theoretical replies and Y_{exp} represents the experimental responses. N is the total number of locations where measurements were made. The validation of the model is deemed valid when $R^2 > 0.7$ and AED 10% [29].

3. One-component Donnan dialysis in the feed compartment

3.1 Counter-ion concentration effect

The effect of counter-ion concentrations in the compartment receiver on the removal of nitrate and nitrite from the feed compartment separately was an important parameter of this investigation. One of the crucial factors influencing the elimination of nitrates and nitrites across the membrane is the counter-ion. Due to its high mobility, chloride appears to be the counter-ion that is employed the most.

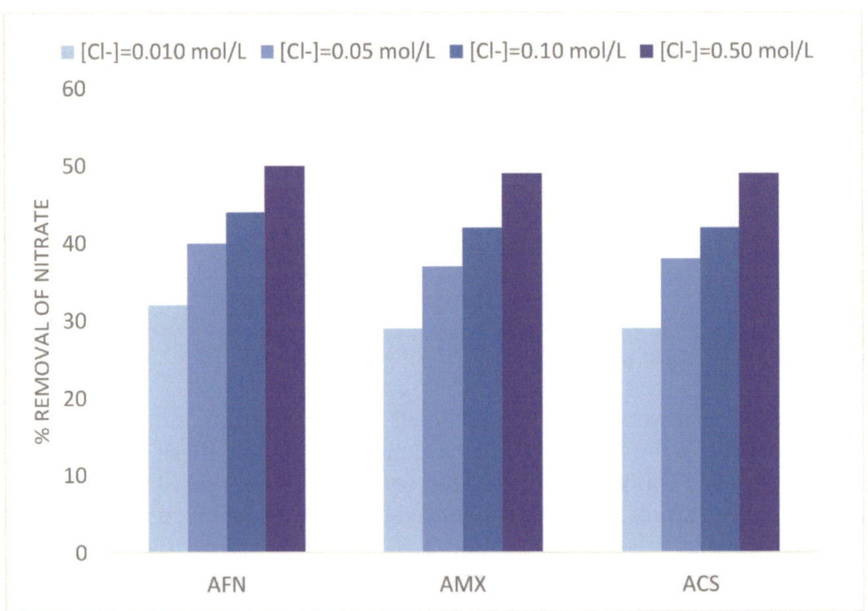

Figure 2.
The counter-ion concentration effect on nitrate removal for the three tested membranes.

We investigated the effect of increasing the concentration of counter-ion Cl^- in the receiver compartment from 0.01 to 0.5 mol./L on the removal of nitrates (100 mg/L) and nitrites (20 mg/L) separately from the feed compartment to the receiver compartment. We have tested three membranes (AFN, AMX, and ACS), with four-hours DD operations. **Figures 2** and **3** show the variation in nitrate and nitrite concentrations at the receiver solution's outlet.

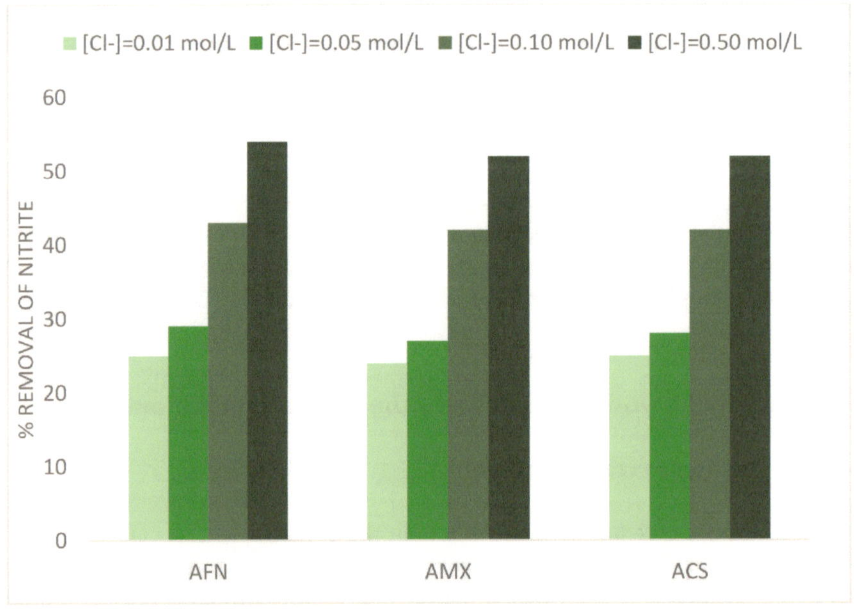

Figure 3.
The counter-ion concentration effect on nitrite removal for the three tested membranes.

For the three membranes, it was observed that increasing the concentration of Cl^- from 0.01 to 0.5 mol/L resulted in a greater removal percentage for nitrate and nitrite. In the receiver compartment, the concentration of nitrate and nitrite was very low when Cl^- was 0.01–0.05 mol/L. As the Cl^-concentration increased from 0.1 to 0.5 mol/L, nitrates and nitrites were greatly removed. To maintain electroneutrality, the cross-ion transfer between Cl^- and nitrates and nitrites improves because the concentration gradient of the counter-ions increases. According to Donnan dialysis with three membranes, an increase in the counter-ion concentration in the receiver compartment is associated with a significant improvement in the removal of nitrates and nitrites in the feed compartment; this is reflected by an increase in the exchange kinetics. Ben Hamouda et al. [13] and Turki et al. [30] have also reported similar conclusions.

3.2 Nitrate and nitrite concentration effect in the feed compartment

Depending on the geographic location, natural waters contained varying levels of nitrate and nitrite. Due to this, nitrate and nitrite concentrations were studied separately in the feed compartment. For this study, the initial concentration of nitrate was varied from 10 to 500 mg/L, and the initial concentration of nitrite was varied from 5 to 100 mg/L. The concentration of counter-ion Cl^- was 0.1 mol/L in the receiver compartment. **Figures 4** and **5** show the variation of nitrate and nitrite concentrations at the receiver solution's outlet.

Figure 4 shows the variation of nitrate concentration from 50 to 500 mg/L, the rate of removal significantly increased for all membranes. Nitrate removal improved when concentration going from 50 to 500 mg/L by 38 to 61% with AFN, 30 to 50% with AMX, and 34 to 50% with ACS. The highest nitrate exchange for chloride ion efficiencies was achieved with the AFN membrane when performing Donnan dialysis.

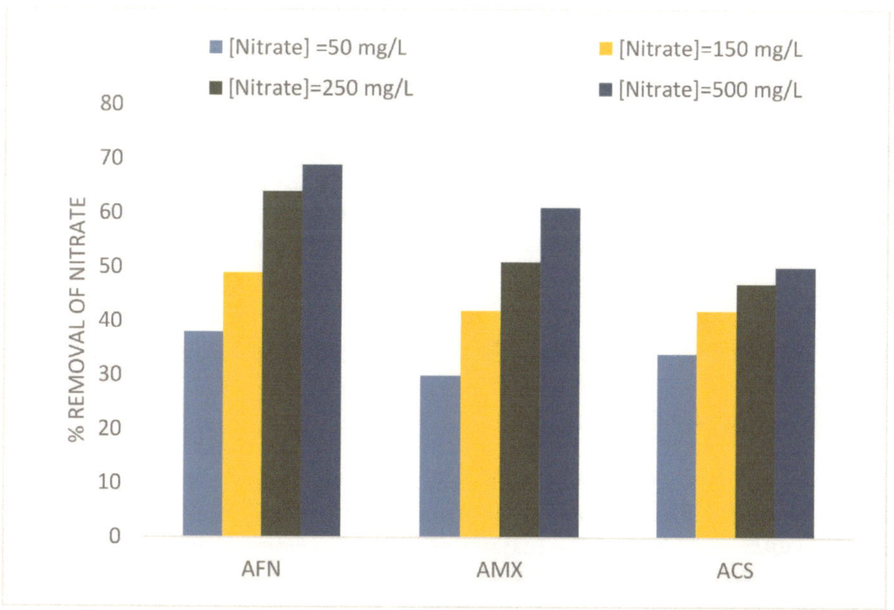

Figure 4.
Nitrate concentration effect in the feed compartment on nitrate removal for the three tested membranes.

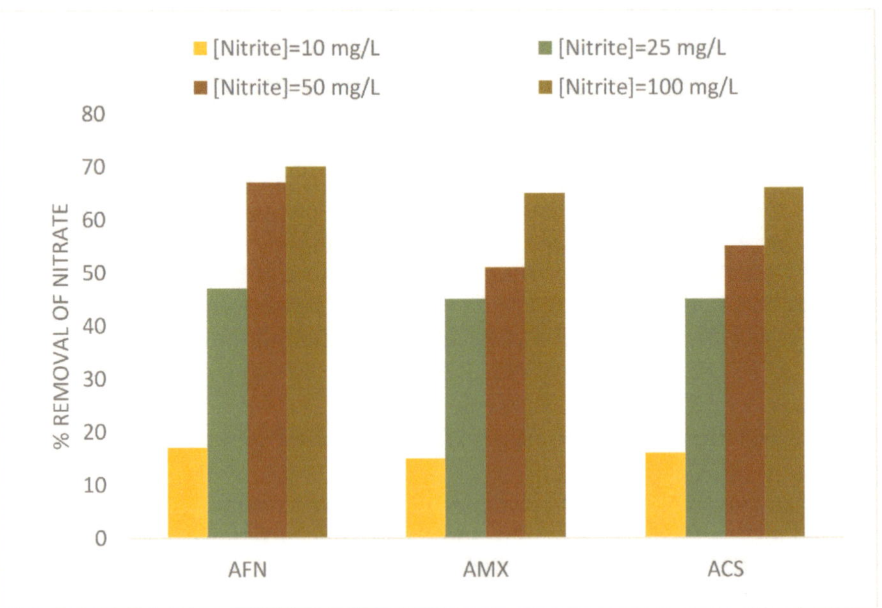

Figure 5.
Nitrite concentration effect in the feed compartment on nitrite removal for the three tested membranes.

Figure 5 shows the variation of nitrite concentration from 10 to 100 mg/L, the rate of removal significantly increased for all membranes. In the feed compartment when the nitrite concentration was the lowest (10 mg/L), the removal for AFN, AMX, and ACS was 17%, 15%, and 16% respectively. The removal was enhanced with an increase in nitrite concentration from 10 to 100 mg/L from 17 to 70% with AFN, 15 to 65% with AMX, and 16 to 66% with ACS. It appears that the AFN membrane removed nitrites most effectively.

We can conclude that the improvement in the removal of nitrate and nitrite was mostly caused by the rise in the initial concentration of nitrate and nitrite, regardless of all membranes AFN, AMX, and ACS. The increase in the concentration gradient of nitrate and nitrite, which increased the chloride ion flux from the receiver compartment to the feed compartment, can be attributed to this [11]. As a result, the cross-ion transfers between Cl^- and nitrates and nitrites are improved.

4. Two components of Donnan dialysis in the feed compartment

4.1 Choice of the membrane

Three membranes, AFN, AMX, and ACS, were examined for the selection of anion-exchange membranes due to the intricacy of the correlation between their qualities given in catalogs and the real performances in Donnan dialysis process. With a counter-ion concentration of 0.5 mol/L, an initial nitrate concentration of 100 mg/L, and a nitrite concentration of 20 mg/L, Donnan dialysis was carried out. **Figure 6** illustrates the simultaneous testing of membranes (AFN, AMX, and ACS) for the removal of both nitrites and nitrates in the same compartment.

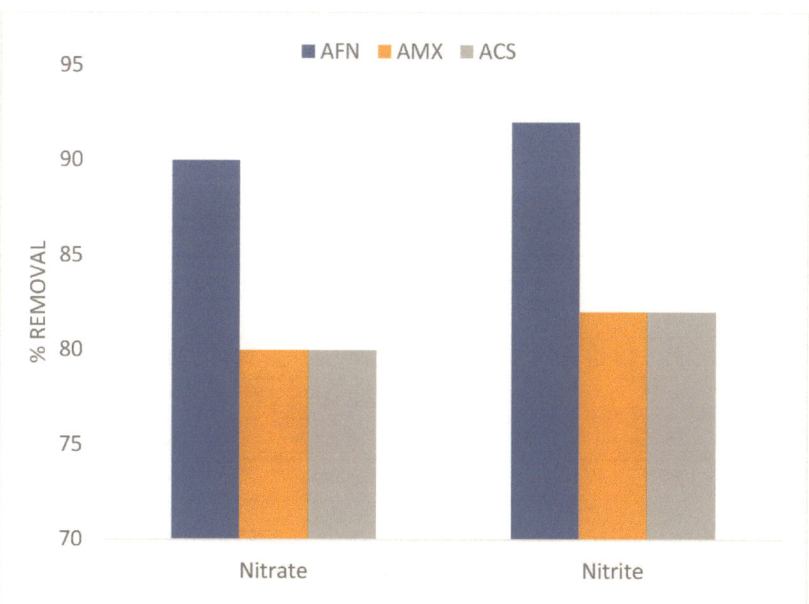

Figure 6.
Choice of the best membrane allowing the highest removal rate for both nitrate and nitrite.

Figure 6 shows the simultaneous elimination of nitrites and nitrates by three membranes (AFN, AMX, and ACS) in the same compartment. Because a higher proportion of nitrates and nitrites ions in the feed compartment increases the overall flow, which in turn causes a higher proportion of the counter-ions to be transported from the feed to the receiver, the presence of nitrite and nitrite in the same compartment improves their elimination.

In comparison to the AMX and ACS membranes, the AFN membrane has the superior performance. The AFN membrane is a macro-porous structure with a high concentration of inorganic groups and a low amount of a cross-linked agent. This membrane has a relatively high ion-exchange capacity and the highest water content [14].

4.2 Doehlert design

The simultaneous removal of nitrate and nitrate was optimized using the membrane AFN and RSM via Doehlert design. The 15 tests of the Doehlert experimental design (**Table 3**) contained three replicates at the center field. Repeated measurements on the same center field can yield almost the same results for our solution, resulting in a significant lack of fit. Replication actually reduces the experimental data's variability, increasing its significance and level of confidence [31].

Using the experimental findings from **Table 3** and the second-order polynomial equations provided by Eqs. (4) and (5), the necessary data were fitted to the equation.

$$\begin{aligned} Y_1 = 87.2 - 1.36\,X_1 + 8.70\,X_2 + 12.35\,X_3 - 2.50\,X_2^2 - 10.40\,X_3^2 + 1.33\,X_1X_2 \\ + 1.33\,X_1X_3 + 1.03\,X_2X_3 \end{aligned} \quad (4)$$

N°	$[NO_2^-]$	$[NO_3^-]$	$[Cl^-]$	Y_1 (%)	Y_2 (%)
1	100	275	0.300	86.3	98.2
2	10	275	0.300	88.1	79.4
3	78	470	0.300	92.7	97.1
4	33	80	0.300	79.1	87.9
5	78	340	0.300	76.3	97.2
6	33	210	0.300	93.2	87.4
7	78	210	0.463	92.9	99.8
8	33	405	0.137	69.9	82.3
9	78	340	0.137	65.7	89.7
10	55	145	0.137	73.7	86.7
11	33	340	0.463	92.7	90.9
12	55	145	0.463	84.2	97.1
13	55	275	0.300	87.2	96.3
14	55	275	0.300	87.2	96.3
15	55	275	0.300	87.2	96.3

Table 3.
Experimental design and results of nitrate and nitrite removal.

$$Y_2 = 96.3 + 9.11\,X_1 - 0.28\,X_2 + 5.94\,X_3 - 7.50\,X_1^2 - 2.50\,X_2^2 - 5.27\,X_3^2 + 0.23\,X_1X_2 + 0.84\,X_1X_3 - 0.04\,X_2X_3 \tag{5}$$

According to obtained results, the coefficients are presented and show that the counter-ion concentration had an important effect (b_3 = 12.35) on the removal of nitrate. The second influenced factor was the nitrate concentration (b_2 = 8.70). Except the concentration of nitrite had a less important effect on the removal of nitrate (b_1 = 1.36). According to obtained results, the coefficients are presented and show that the nitrite concentration had an important effect (b_1 = 9.11) on the removal of nitrite. The second influenced factor was the counter-ion concentration (b_3 = 5.94). Nevertheless, the concentration of nitrate had a less important effect on the removal of nitrate (b_2 = 0.28).

The multiple polynomial model coefficients (Eqs. (4) and (5)), which reflected the effects and interactions of the many components under investigation were identified. A check of the relative importance of various coefficients in the experimental region under investigation is possible thanks to the Pareto analysis (**Figure** 7). The following relation (Eq. (6)) allows to calculate this analysis:

$$P_i = \left(\frac{b_i^2}{\sum b_i^2}\right)^2 \times 100 \tag{6}$$

The concentration of chloride and nitrate both have a positive effect on the studied response, meaning that increasing their concentrations leads to improved nitrate removal. Their contributions to the studied response were only 30.5% for chloride

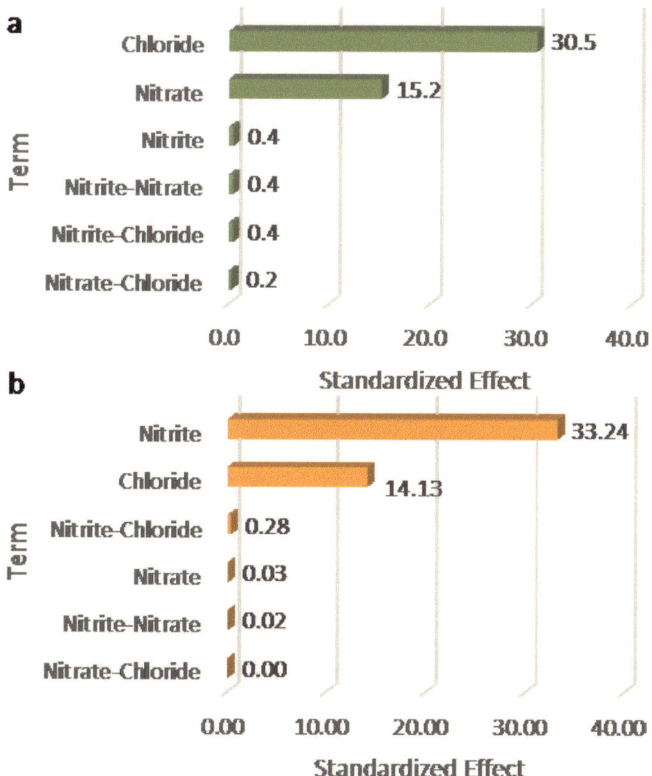

Figure 7.
Pareto analysis for nitrate (a) and nitrite (b) removal.

concentration and 15.2% for nitrate concentration. Thus, the removal can be considerably influenced by two parameters: chloride concentration and nitrate concentration. However, the other interactions have a negligible effect; they represent only 1.0% of the studied response. The concentration of nitrites and the concentration of chloride are two factors that affect nitrate elimination. The concentration of nitrite has a positive effect, which translates to improved elimination via Donnan Dialysis as nitrite concentration rises. This improvement was attributed to the rise in the concentration gradient of nitrates, which enhanced the chloride ion flux from the receiver to the feed solution, hence the cross-ion transfer between Cl^- and nitrate improves to maintain the electroneutrality.

Two factors positively impact the studied response, indicating that an increase in these factors leads to an improvement in nitrite removal. Their contributions to the studied response were only 33.24% for nitrite concentration and 14.13% for chloride concentration. The nitrate concentration has a negative effect, meaning that an increase in nitrate concentration leads to a decrease in nitrite removal. However, the interaction between nitrite and chloride has a negligible effect of 0.28%. Similarly, the other interactions also have a negligible effect, accounting for only 0.3% of the studied response.

The concentration of nitrites and the concentration of chloride are two factors that affect nitrite elimination. The concentration of nitrite has a positive effect, which translates to improved elimination via Donnan Dialysis as nitrite concentration rises.

This improvement was attributed to the rise in the concentration gradient of nitrates and nitrites, which enhanced the chloride ion flux from the receiver to the

Source model	Sum of square	Degree of freedom	Mean of square	F-value	P-value
NO_3^-					
Regression	1149.0	9	1272.10	86.8865	0.0005
Residual	7.3205	5	1.464		
Total	1152.2	14			
NO_2^-					
Regression	1806.89	9	61.93	223.9885	0.0001
Residual	1.3825	5	0.2765		
Total	558.77	14			

Table 4.
Variance analysis of nitrate and nitrite removal.

feed solution, hence the cross-ion transfer between Cl^- and nitrite improves to maintain the electroneutrality.

Using the regression coefficient (R^2) and the percentage of absolute errors of deviation (AED), the validity of the model was evaluated. According to Y_1, the regression coefficient is 0.999, while according to Y_2, it is 0.997. However, the percentage absolute error of deviation, AED (%) = 0.262%, was less than 10% for nitrate removal, and AED (%) = 0.311% for nitrite removal. These confirm the validation of the models suggesting that the model is suitable to describe the removal of nitrate and nitrite.

Analysis of variance (ANOVA) was carried out to support the models' suitability. The results are shown in **Table 4** in this regard. The F-ratio is the mean square error, and the p-value is the ratio of the mean square effect.

The P-value has been used to identify the effects that are statistically significant. The P-value is crucial since it is close to zero, which denotes the significance of the data. The Fischer value ($F_{0.05,\ 1,\ 1.16}$) for 5% error, 1 degree of freedom, and 16 factorial testing is 4.77, according to the Fischer table. Due to the fact that every effect's value is higher than 4.77, it appears that they are all important. At a level of 5%, the experimental model's Fischer value is significantly greater than the crucial F value. The model is therefore regarded as statistically significant.

The software NemrodW®s desirability function produced the optimal conditions. The optimum values for each factor are therefore 82 mg/L for nitrite concentration, 406 mg/L for nitrate concentration, and 0.412 mol/L for counter-ion concentration. The maximal elimination of nitrates (95.5%) and nitrites (100%) was achieved under these conditions. In order to confirm the accuracy of the predictions, the experiment was replicated three times under optimal conditions. Since the coefficient of repeatability was less than 1%, it can be said that Donnan dialysis's simultaneous removal of nitrates and nitrites is repeatable. The experimental results' statistical analysis revealed that the analysis has a normal distribution.

5. Application on real water

Simultaneous removal of nitrate and nitrite from real water was performed to ensure the feasibility and effectiveness of Donnan dialysis. We used water from the eastern side of the Tunisian Cape Bon, a region well-known for its intensive

Parameters	Concentration (mg/L) or value
Fluoride	0.8
Chloride	528.9
Nitrate	98.4
Nitrite	8.9
Phosphate	5.2
Sulfate	127.5
pH	7.8

Table 5.
Composition of the groundwater from Menzel Bouzelfa.

cultivation of vegetables and fruit, mainly citrus. The large quantities of chemical fertilizers have caused real problems with the quality of both surface and groundwater. **Table 5** shows the anion content of water taken from a well in the town of Menzel Bouzelfa. We notice the high contents of nitrates and nitrites which are respectively of the order of 100 and 10 mg/L.

We used the experimental parameters resulting from the optimization of this process for similar nitrate and nitrite contents, namely a chloride concentration of 0.4 mol/L and a removal time of 4 hours under agitation. We also tested the three membranes AFN, AMX, and ACS. The results obtained are given in **Figure 8**. It can be seen that the most efficient membrane is AFN and that the removal rates for this membrane are around 98% for nitrates and 85% for nitrites. These results are perfectly consistent with theoretical predictions and confirm that Donnan dialysis remains very efficient for the removal of these two anions.

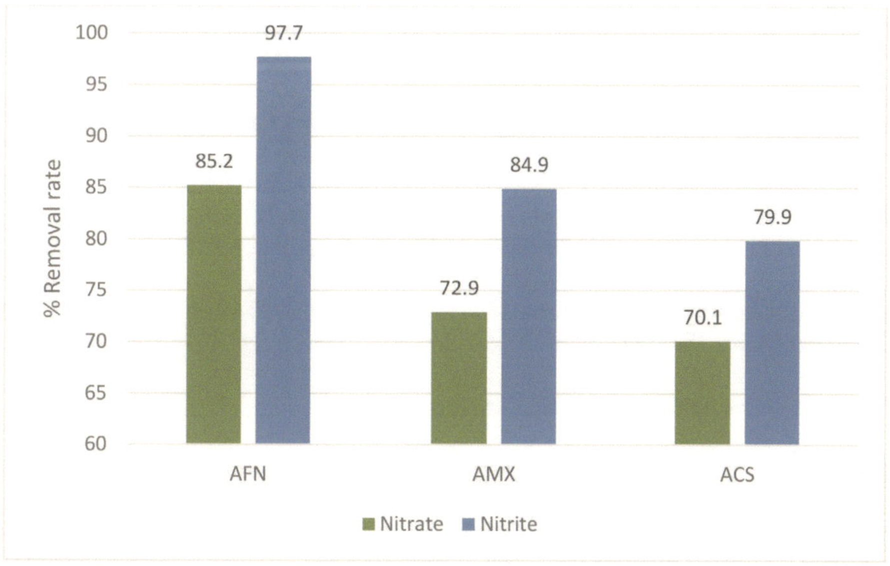

Figure 8.
Simultaneous removal of nitrate and nitrite from the real water of Menzel Bouzelfa—Tunisia.

6. Conclusion

In order to clearly depict the impact of every factor on optimum conditions and to determine the optimal parameters for simultaneous removal of nitrate and nitrite by Donnan dialysis with a small number of experiments, RSM according to the Doehlert matrix was used. The RSM is highly efficient and makes it easier to achieve the optimum conditions while accounting for interactions between experimental parameters. These conditions were 82 mg/L for the concentration of nitrite, 406 mg/L for the concentration of nitrate, and 0.412 mol/L for the concentration of chloride, which were the best for the simultaneous removal of nitrates (95.5%) and nitrites (100%) through the AFN membrane. The removal effectiveness of the simultaneous removal of nitrite and nitrate by DD as well as the optimization of process factors for maximal removal were successfully determined by the RSM approach. Finally, the application of the DD for treating real water (from the region of Menzel bouzelfa) to remove nitrate and nitrite allowed us to confirm that AFN is the best membrane and that the elimination rates remain very high and close to those found in the theoretical study.

Acknowledgements

The researchers would like to thank the ICMPE/University Paris-Est Créteil for funding the publication of this project.

Author details

Ikhlass Marzouk Trifi[1*], Beyram Trifi[2] and Lasâad Dammak[3]

1 Desalination and Water Treatment Laboratory (LDTE), Faculty of Sciences of Tunis, University of Tunis El Manar Tunis, Tunisia

2 Materials, Processing and Analysis Laboratory (LMTA), National Institute for Research and Physico-chemical Analysis (INRAP), Biotechpole Sidi Thabet, Tunisia

3 Institute of Chemistry and Materials Paris-Est (ICMPE), Paris-Est University, UMR 7182, CNRS, Thiais, France

*Address all correspondence to: ikhlassmarzouk@gmail.com

References

[1] Aslan S, Turkman A. Nitrate and pesticides removal from contaminated water using biodenitrification reactor. Process Biochemistry. 2006;**41**(4): 882-886. DOI: 10.1016/j.procbio. 2005.11.004

[2] Altintas O, Tor A, Cengeloglu Y, Ersoz M. Removal of nitrate from the aqueous phase by Donnan dialysis. Desalination. 2009;**239**(1–3):276-282. DOI: 10.1016/j.desal.2008.03.024

[3] Calderer M, Jubany I, Pérez R, Martí V, de Pablo J. Modelling enhanced groundwater denitrification in batch micrococosm tests. Chemical Engineering Journal. 2010;**165**(1):2-9. DOI: 10.1016/j.cej.2010.08.042

[4] WHO. Health hazards from nitrate in drinking-water. Report on a WHO Meeting: Copenhagen, 5-9 March 1984, Conference Report, WHO. Regional Office for Europe-Copenhagen, DK, EURO. Available from: https://www.ircwash.org/resources/health-hazards-nitrates-drinking-water-report-who-meeting-copenhagen-5-9-march-1984

[5] Lacasa E, Cañizares P, Sáez C, Fernández FJ, Rodrigo MA. Removal of nitrates from groundwater by electrocoagulation. Chemical Engineering Journal. 2011;**171**(3): 1012-1017. DOI: 10.1016/j. cej.2011.04.053

[6] Rossi F, Motta O, Matrella S, Proto A, Vigliotta G. Nitrate removal from wastewater through biological denitrification with OGA 24 in a batch reactor. Water. 2015;7:51-62. DOI: 10.3390/w7010051

[7] Taziki M, Ahmadzadeh H, Murry MA, Lyon SR. Nitrate and nitrite removal from wastewater using algae. Current Biotechnology. 2015;**4**:426-440. DOI: 10.2174/2211550104666150828193607

[8] Ye J, Liu SQ, Liu WX, Meng ZD, Luo L, Chen F, et al. Photocatalytic simultaneous removal of nitrite and ammonia via a zinc ferrite/activated carbon hybrid catalyst under UV–visible irradiation. ACS Omega. 2019;**4**: 6411-6420. DOI: 10.1021/ acsomega.8b00677

[9] Strathmann H. Ion-Exchange Membrane Separation Processes. 1st ed. Vol. IX. Amsterdam-Boston: Elsevier; 2004

[10] Barros KS, Carvalheira M, Marreiros BC, MAM R, Crespo JG, Pérez-Herranz V, et al. Donnan dialysis for recovering ammonium from fermentation solutions rich in volatile fatty acids. Membranes. 2023;**13**:347. DOI: 10.3390/membranes13030347

[11] Wisniewski J, Rozanska A, Winnicki T. Removal of troublesome anions from water by means of Donnan dialysis. Desalination. 2005;**182**:339-346. DOI: 10.1016/j.desal.2005.02.032

[12] Turek M, Bandura-Zalska B, Dydo P. Boron removal by Donnan dialysis. Desalination Water Treat. 2009;**10**: 53-59. DOI: 10.5004/dwt.2009.711

[13] Ben Hamouda S, Touati K, Ben AM. Donnan dialysis as membrane process for nitrate removal from drinking water: Membrane structure effect. Arabian Journal of Chemistry. 2017;**10**: S287-S292. DOI: 10.1016/j.arabjc.2012. 07.035

[14] Hichour M, Persin F, Sandeaux J, Gavach C. Fluoride removal from waters

by Donnan dialysis. Separation and Purification Technology. 2000;**18**: 1-11. DOI: 10.1016/S1383-5866(99) 00042-8

[15] Dieye A, Larchet C, Auclair B, Mar-Diop C. Elimination des fluorures par la dialyse ionique croisée. European Polymer Journal. 1998;**34**:67-75. DOI: 10.1016/S0014-3057(97)00079-7

[16] Marzouk I, Dammak L, Chaabane L, Hamrouni B. Optimization of chromium (VI) removal by Donnan dialysis. American Journal of Analytical Chemistry. 2013;**4**:306-313. DOI: 10.4236/ajac.2013.46039

[17] Marzouk I, Chaabane L, Dammak L, Hamrouni B. Application of Donnan dialysis coupled to adsorption onto activated alumina for chromium (VI) removal. American Journal of Analytical Chemistry. 2013;**4**:420-425. DOI: 10.4236/ajac.2013.48052

[18] Trifi IM, Trifi B, Djamel A, Hamrouni B. Simultaneous removal of nitrate and nitrite by Donnan dialysis. Environmental Engineering and Management Journal. 2021;**20**:6973-6983. DOI: 10.30638/eemj.2021.090

[19] Breytus A, Hasson D, Semiat R, Shemer H. Removal of nitrate from groundwater by Donnan dialysis. Journal of Water Process Engineering. 2020;**34**: 101157-101164. DOI: 10.1016/j.jwpe. 2020.101157

[20] Imandi SB, Bandaru VV, Somalanka SR, Garapati HR. Optimisation of medium constituents for the production of citric acid from by product glycerol using Doehlert experimental design. Enzyme and Microbial Technology. 2007;**40**: 1367-1372. DOI: 10.1016/j.enzmictec. 2006.10.012

[21] French Standard NF X 45-200, Membranes Polymers Échangeuse d'Ions. Paris, France: AFNOR; 1995

[22] Marzouk Trifi I, Trifi B, Ben Ayed S, Hamrouni B. Removal of phosphate by Donnan dialysis coupled to adsorption onto alginate calcium beads. Water Science and Technology. 2019;**80**: 117-125. DOI: 10.2166/wst.2019.256

[23] Rodier J. L'analyse de l'eau. 9th ed. Paris: Edition Dunod; 2009

[24] Goupy J, Greighton L. Introduction to Design of Experiments. 3rd ed. Paris: Dunod; 2006

[25] Hannachi A, Hammami S, Raouafi N, Maghraoui-Meherzi H. Preparation of manganese sulfide (MnS) thin films by chemical bath deposition: Application of the experimental design methodology. Journal of Alloys and Compounds. 2016; **663**:507-515. DOI: 10.1016/j.jallcom. 2015.11.058

[26] Hosni N, Bouaniza N, Selmi W, Assili K, Maghraoui-Meherzi H. Synthesis and physico-chemical investigations of $AgSbS_2$ thin films using Doehlert design and under DFT framework. Journal of Alloys and Compounds. 2019;**778**:913-923. DOI: 10.1016/j.jallcom.2018.11.072

[27] Hosni N, Zehani K, Djebali K, Mazaleyrat F, Bessais L, Maghraoui-Meherzi H. Experimental design approach for the synthesis of 3D-$CoFe_2O_4$ nanoflowers thin films by low-cost process. Materials Chemistry and Physics. 2020;**255**:123493. DOI: 10.1016/ j.matchemphys.2020.123493

[28] Ouejhani A, Hellal F, Dachraoui M, Lalleve G, Fauvarque JF. Application of Doehlert matrix to the study of

electrochemical oxidation of Cr(III) to Cr(VI) in order to recover chromium from wastewater tanning bath. Journal of Hazardous Materials. 2008;**157**: 423-431. DOI: 10.1016/j.jhazmat. 2008.01.046

[29] Nde BD, Abi CF, Tenin D, Kapseu C, Tchiegang C. Optimisation of the cooking process of sheanut kernels (Vitellaria paradoxa Gaertn.) using the Doehlert experimental design. Food Bioprocess Technology. 2012;**5**:108-117. DOI: 10.1007/s11947-009-0274-z

[30] Turki T, Hamdi R, Tlili M, Ben AM. Donnan dialysis removal of nitrate from water: Effects of process parameters. American Journal of Analytical Chemistry. 2015;**6**:569-576. DOI: 10.4236/ajac.2015.66055

[31] Papaneophytou C. Design of Experiments As a Tool for Optimization in Recombinant Protein Biotechnology: From Constructs to Crystals. Molecular Biotechnology. 2019;**61**:873-891. DOI: 10.1007/s12033-019-00218-x